# FLAMMABILITY

# OF POLYMERIC

# MATERIALS

# FLAMMABILITY

# OF POLYMERIC

# MATERIALS

## Editor

NOVA SCIENCE PUBLISHERS, INC.
COMMACK

Art Director: Maria Ester Hawrys
Assistant Director: Elenor Kallberg
Graphics:  Kerri Pfister, Erika Cassutti and
                 Barbara Minerd
Manuscript Coordinator: Sharyn Schweidel
Book Production: Tammy Sauter, and Benjamin Fung
Circulation:  Irene Kwartiroff and Annette Hellinger

*Library of Congress Cataloging-in-Publication Data*

Flammability of polymeric materials / G. E. Zaikov (ed.).
     p. cm.
Includes bibliographical references and index.
ISBN 1-56072-256-8 :
1. Polymers--Fire testing.  I. Zaikov, Gannadii
Efremovich.
  TH9446.5.P65F58   1995                              95-24682
  628.9'222--dc20                                         CIP

© 1996 Nova Science Publishers, Inc.
   6080 Jericho Turnpike, Suite 207
   Commack, New York  11725
   Tele. 516-499-3103  Fax 516-499-3146
   E Mail Novasci1@aol.com



*Printed in the United States of America*

# CONTENTS

# INTRODUCTION

One from the most important problems today in the field of physics and chemistry at polymers and polymeric matrices is the problem of flammability and flame retardants. More and more polymer materials are being used for different purposes and thus there are more and more possible fire accidents. During fires, more than half of the people die from toxic products from combustible materials. For this reason, it is very important to use neither toxic polymers nor toxic derivatives for polymers. I would like to offer here two phrases which illustrate this problem: "There are two possibilities - either mankind will make less smoke to be on Earth or the smoke will make less men to live" (G. Burton), and "Our globe resembles a car running alone in space with the exhaust pipe revealed into the passengers' salon. The more force we apply to the accelerator, the higher the probability that we will poison the observer as well as the passengers" (Jacques Ives Cousteau). The purpose of this volume is research for solving this problem.

G.F. Zaikov, Institute of Biochemical Physics

# FLAME RETARDANCY OF CELLULAR POLYMERIC MATERIALS


**R.M. Aseeva and G.E. Zaikov**
*Institute of Chemical Physics,*
*Russian Academy of Sciences,*
*4, Kosygin Str., 117334, Moscow, Russia*



The combustion behavior of cellular polymeric materials is reviewed. The correlations between the structure and thermophysical properties of cellular polymers are discussed. The estimation tests for potential fire hazard of cellular polymeric materials are considered. The approaches to flame retardancy of rigid foams based on reactionable polyfunctional olygomers (phenolformaldehyde and urea formaldehyde ones, polyurethane) are demonstrated.

**Key words:** Cellular polymer, foam, flammability, flame retardancy


## 1. INTRODUCTION

Cellular polymeric materials (CPM) are notable for the unique combination of technically valuable properties. With relatively small weight they have sufficiently high strength and are characterized by thermal insulating or soundabsorbing properties. Floatage, filtration capacity or selective permeability, the ability to resist a considerable temperature changes are some of other interesting characteristics of CPM.

High qualities of cellular polymers together with low polymer containing stipulated the increased interest to this kind of synthetic products.

At present time the volume of world production and the consumption of CPM consist on the average about 10% of the whole plastic production[1]. However, for some polymers wide-spread in industrial countries (polyurethanes, for example) the part of CPM reaches 50-80%[2]. The production of some CPM is developing faster in comparison with other types of plastics. It is because CPM are high-efficient products, which permit to economize raw materials and energy resources.

According to the main directions in economical and social development of Russia for the period up to 2000 in the area of energetic and complex national programs a considerable increase of CPM production was foreseen. The number of produced CPM was planned to be broadened and their exploitational and technico-economical indices to be improved. The volume of CPM consumption by just Russian building industry in 1990 had to be over 5 million $m^3$. However, at present time the considerable decrease of the production of all kind plastics is observed in the region in connection with political and economical crisis, the USSR disintegration into a number of new independent states.

Meanwhile the importance of CPM for Russia may be demonstrated on the example with the application of heat communications. About 200 thousand km of heatnets with pipes of diameter up to 600 mm and about 20 thousand km of pipe lines with diameter of 600-1400 mm are in exploitation in Russia. The quantity of heat energy by means of hot water or steam, transported by these communications, is over 9.6 billion GJ/year, i.e. it is about 70% of total consumption of this kind of energy.

The losses of heat reaches up to 10% of transported energy, because of great length of arterial and separating heat communications, including ones, situated in regions with hard climatic conditions. According to the estimations heat losses of transported energy in 1984 were equal to one of about 60 million tons of conventional fuel.

The use of high efficient thermal insulating CPM instead of traditional mineral-wadding articles makes not only essential decrease of losses possible, but also increase the effectiveness of labour and improves sanitary-hygienic conditions during assembly building works.

Furniture industry, transport, the production of packing materials, freezing techniques, etc. are th emain customers of CPM besides building industry.

The most of celllar polymers are flammable. The requirements to fire safety of polymer materials are increased. The dvelopment of fire safe materials is one of the most important social and economical problems.

Conflagrations cause a great damage to the society, irreplaceable in some cases (people death, destruction of culture memorials). According to data of the World Centre of fire statistics the society losses from fires in 1977-1980 were as0.4 up to 0.8% from national budgets for different countries[3]. In 1982-1983 they were from 0.1 to 0.48%[4] of ones. The decrease of the damage is in straight dependence on system of measues for prevention and struggle against fires. Objective estimation of fire hazard of materials, their rational using, development of nontoxic flame retardant, low smoke articles are the important components of this system. Due to specific structure features of the heterophasic polymer systems, CPM differ essentially from monoithic, nonfoamed materials by their physical properties [5-7]. It induces quite different reaction of CPM on the action of fire or high temperatures. Chemical nature of polymer or chemical composition of CPM play the first order role in the consideration of CPM fire hazard characteristics. However, physical or morphological structure of CPM also affect the appearance and the development of combustion process.

In the present paper the combustion behavior of CPM is reviewed. The correlation between the structure and thermal physical properties of cellular polymers are discussed. The tests for potential fire hazard estimation of CPM are considered. Rigid foams based on reactionable polyfunctional olygomers (phenol formaldehyde and urea formaldehyde ones, polyurethanes) were the main objects of the study. The approaches to flame regardancy of these CPM are demonstrated.

## 2. ON STRUCTURAL FEATURES AND SOME PROPERTIES OF CELLULAR POLYMERS

Cellular polymers could be considered as compositional materials, containin specific filler- air or another gas. Gas inclusions impart to polymers a combination of new properties, inherent in both polymer and filling gas.

It is conventional to divide cellular polymers into foams and poreplastics according to a kind of gas inclusions (cell, pores), viz. if they are isolated or connected. However, very often cellular polymers are being produced with mixed structure, with a definite ratio of closed and opened cells. Now technology of the production for other types of gas-filled polymers has been worked out, such as integral (structured) foams, syntactic and multilayere (laminate) foams. CPMs may be produced as foam films or foam fibers, mixed foams, reinforced materials or multilayer construction.

To describe the properties of CPM it is not enough to consider only form and size of cells (pores), type of connection of cells. It is necessary to pay attention to the structure of intercellular space, being polymer matrix itself.

According to modern conceptions [5-7] the structure of different cellular polymers, fully defining their properties, includes 6 levels of organization: 1 - molecular chemical structure of polymer; 2 - conformational structure of macromolecules; 3 - submolecular structure of polymer matrix; 4 - spacial structure of gas cells and polymer matrix in the intercellular space; 5 - structure and localization of microcells in walls and crosspieces of macrocells; 6 - subcellular structure, characterizing the distrubition of gas-structural elements in the bulk of the material.

The first and second of CPM organization levels are responsible for all chemical processes of matrix transformation into combustion products. Morphological structure organization levels define macroscopic properties of CPM including their reaction on flame and high temperatures at fire situations.

The key to the realization of CPM fire safety lies in the understanding of the interconnection between chemical and morphological organizations, on the one hand, and macroscopic properties of gas-filled polymers, especially those, influencing on heat and mass transfer.

The conception of "gas-structural element" suggested by A.A. Berlin and F.A. Shutov[5] is for the benefit of quantitative description of CPM morphological structure. Gas-structural element is a spacial one, consisting of gas cavity - cell, walls and ribs (struts), and tangles of crosspieces either. This spacial element of solid and gas phases is reiterated in foam material with a definite period and regulation, degree of intercellular space closing, thus creating the macrostructure of material.

For quantitative characterizing of CPM macrostructure a number of parameters are being used. They are: apparent density (bulk weight); porosity (gas filling), including active one, which participates in gas filtration movement; number and ratio of open and closed cells; form, size and cell distribution by size and form either; distribution of polymer matrix in walls and struts; specific surface of foams, etc.

Porosity, $\Phi$, is determined by the ratio of volume, occupied by gas pores (cell), $V_g$, and a total volume of system, $V_g$: $\Phi = V_g/V_0$. The relative gas volume in the foam could be defined as:

$$V_g = 100(\rho_s - \rho)/\rho_s,$$

where $\rho_s$ and $\rho$ are density of nonfoamed polymer and apparent density of CPM, respectively.

Integral CPM are characterized additionally by the distribution of apparent density through bulk material. For CPM, consisting of polymer matrix and gas, volume fraction of polymer, $v_s$, and gas, $v_g$, are: $v_s = \rho/\rho_s$; $v_g = 1 - v_s$.

To a definite value of porosity a great number of structural forms with different gas-structural element position can be corresponded.

According to the condition of the production CPMs contain isolated and communicated cells in their structure in defined proportion. Optical and scanning electron microscopy are usually used for structural investigations of CPM. However, the precisional determination of the quota of opened cells is rather complicated due to the planar character of pictures. That is why sorptional methods are used in practice. For example, method of moisture absorption or volumetrical displacement method[8].

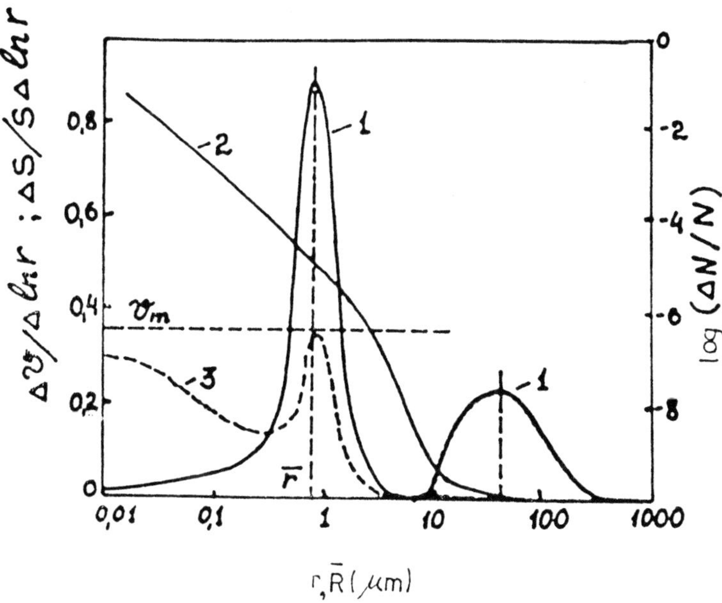

**Figure 1.**   Distribution curves of morphological parameters for PU foam (r=42 kg/m³) 1 - differential distribution of micro and macrocells depend on sizes (DV/Dlnr); 2 - distribution of relative number of cells (lg DNi/N); 3 - distribution of specific surface (DS/SDlnN).

The idealized forms (monodispersed spheres, hexagons, rhombic dodecahedrons, stretched pentagonal dodecahedrons, capillaries, etc.) were suggested as models of morphologic structure of gas-filled polymers[9]. The deviation from idealized structure is reflected in existence of anisotropic forms, connected with a direction of foaming.

Sizes of cells and their distribution in foam structure are estimated by statistic treatment of data, obtained with the help of straight methods of structure investigation.

F.A. Shutov [5,6,10] was the first to find a biomodal distribution of cell sizes in the structure of rigid foams based on reactionable olygomers (Figure 1). Two types of cells are present in the structure of these CPMs. They are micro and macrocells with sizes of 0,5-1 μm and 60-70 μm respectively. Microcells present the main portion (up to 99%)

of total number of cells. However, the contribution of microcells to the general porosity is rather small: about 5-10%.

Unlike macrocells microcells have spheric form. They can be isolated or connected. Microcells are located in struts and cell walls. The calculations showed that the thickness of macrocell walls is usually by two decimal degrees lower than their diameter. Meanwhile maximum thickness of microcell walls is nearly equal to their radius. Linear sizes of microcells do not practically depend on apparent density of CPM.

Large specific surface, both by volume, $S_v$, and mass, $S_m(S_m=S_v/\rho)$, is another important feature of CPM with bimodal distribution of cell sizes. The specific surface of rigid foams can exceed 100 $m^2/g$. For CPM based on reactionable olygomers microcell formation mechanism is connected with the formation of three dimensional net polymer structure while foaming. It is possible to regulate macro- and microcell structure of foams and also their properties by the action on temperature-time recycles of foaming and polymer matrix hardening. The character of polymer distribution in gas-structural elements of CPM affect both transportational and mechanical or other properties of foams.[11]

Let us consider heat transfer in CPM. The mechanism of heat transfer includes one by thermal conductivity through solid and gas, by convection through gas phase and also by radiation. Considering these components independent, CPM thermal conductivity can be presented as: $\lambda=\lambda_s+\lambda_g+\lambda_c+\lambda_r$.

For a large number of CPM with porosity up to 97% solid phase content by volume is small. This part cannot be ignored as the thermal conductivity of polymer matrix exceeds one of gas in cells. Nonuniform polymer distribution in gas-structural elements and their anisotropy stipulate different contribution of heat transfer through foam in longitudinal and perpendicular directions according to foaming one[12]. Gas in cells gives the main contribution into heat transfer in foam. For CPM with closed cellular structure a gas used as a blowing agent is the main component of gas phase. In open cellular CPM gas phase consists of air.

Thermal conductivity of gas phase in CPM, having both types of cells, can be calculated by the equation: $\lambda_g=v_{cl}\lambda_{mix}+(1-v_{cl})\lambda_{air}$, where $v_{cl}$ is voluminal part of closed cell; $\lambda_{mix}$ is thermal conductivity of gas mixture in closed cells; $\lambda_{air}$ is air thermal conductivity. Thermal conductivity of mixture of two gases (freon and air, as example):

$\lambda_{mix}=\lambda_1{}^{v_1}\lambda_2{}^{v_2}$, where $v_1$ and $v_2$ are mole parts of gases. It is known that thermal conductivity of gas decreases with its molecular mass growth.

The contribution of convection to heat transfer through CPM becomes noticeable at average cell diameter of 4 mm and higher[13]. This contribution is very small for natural convection and is usually neglected. For porous solids with high gas permeability in presence of filtrational flows one should not neglect conventional contribution. It is connected with gas thermal conductivity and fluid dynamic parameter by empyric relationship[14]: $\lambda_c=1.10^{-2}\lambda_g Re_{eff}$; $Re_{eff}=V_f d_{eff}/v$, where $Re_{eff}$ is Reynolds number; $V_f$ - gas flow rate; $d_{eff}$ - effective pore size; $v$ is kinematic gas viscosity.

Let us consider heat transfer through CPM by radiation. The radiation contribution can be noticeable. The equation was suggested for calculation of radiation contribution in CPM heat transfer in the supposition that all the polymer is located in cell walls[15]: $\lambda_r=4\sigma T^3 L/[1+(L/L_g)(1/T_N-1)]$, where $T_N$ is the portion of radiation energy transmitted through cell wall; $T$ is average temperature of layer; $\sigma$ is Stephan-Boltsmann constant; $L$ and $L_g$ are wall thickness and cell size. $T_N$ can be estimated by properties of polymer matrix: $T_N=(1-r)/(1-2t)[[\{(1-r)t/(1+2t)\}+(1-t)/2]$, where $r$ and $t$ are coefficients of reflection and transmission of radiation energy by polymer.

More accurate estimation of radiation contribution in heat transfer with account for distribution of polymer matrix in struts with tangles of cells was suggested in work[16]. Effective thermal conductivity of CPM with closed cellular structure (as example, PU foams, including dodecahedron elements) is given by the relationship:

$$\lambda=\lambda_g[(2/3)-v_{s,t}/3]+(1-\Phi)\lambda_s+16\ \sigma T^3\ l_r/3K$$

The equation reflects the contributions to heat transfer by conduction through gas and solid phases of CPM and by radiation. K represents average extinction coefficient of grey solids in UV region of spectrum. $v_{s,t}$ is voluminal portion of polymer in struts. Accepted numeral coefficients calculate that 1/3 of struts is oriented in the heat flow direction.

Analysis of factors, which permit to decrease all components of heat transfer, made it possible to recieve PU foams with a very low effective thermal conductivity (0.014-0.015 W/m.K)[17].

The influence of apparent density on thermophysical properties of foams based on reactionable olygomers used in the work are described by next regression equations[18] for phenolic foams with $\rho$ of 40-1200 kg/m$^3$:

$$\lambda=2{,}85.10^{-2}+1{,}84.10^{-4}\rho,\ W/m.K;\ \text{thermal diffusivity}$$
$$a=(240/\rho)+0{,}7;\ (a.10^7),\ m^2/s$$

for polyurethane foams with $\rho$ to 325 kg/m$^3$:

$$\lambda=2{,}85.10^{-2}+0{,}85.10^{-4}\rho;\ a=(218/\rho)+0{,}3;\ (a.10^7),\ m^2/s.$$

If the apparent density $\rho$, volume portion of open cell $v_{open}$, and volume portion of closed cell with freon, $v_{fr}$, are taken into account, the next regression equation for calculation of thermal conductivity are used[19]:

$$\lambda=34,76-0,142\rho-0,08v_{open}-0,10v_{fr}+2,5.10^{-3}\rho^2+7,3.10^{-4}v^2_{open}+5.10^{-4}\rho$$
$$v_{open}-1,5.10^{-3}\rho\ v_{fr}+1,6.10^{-3}v_{open}\ v_{fr};\ mW/m.K.$$

# 3. TESTING OF FIRE HAZARD INDICES OF CELLULAR POLYMERIC MATERIALS

There are a considerable number of tests to study flammability and fire hazard of CPM. The test methods are divided on bench-scale (to 120 cm), large-scale (above 120 cm) and natural ones by principle of size correlation for testing sample and real article[20]. Although there are some difficulties in relating the results of bench-scale flammability tests with real fire behaviour for CPM, small and middle-scale tests continue to be of value for a number of investigation purposes.

Some general bench-scale methods for fire hazard testing of polymeric materials, used in Russia, correspond to state standard of GOST 12.01-44-89[21]. The standard includes the methods for the estimation of ignitability indices, flame spread, gas toxicity and smoke formation at combustion or pyrolysis of polymeric materials. The standard has been developed with account to different ISO tests. In the present work the next tests have been used in accordance to GOST 12.01-44-84:
limiting oxygen index (correpond to ISO 4589); ignition and self-ignition temperatures (analogous to ASTM D 1929); flame spread index and ceramic tube index (CT index).

CT index represents the ratio of the heat released as a result of polymer combustion to the heat required for the sample ignition. At CT index values less of 1 the materials are considered hard combustible. At CT index more of 1 materials are combustible. Among combustible ones the group of hard ignitable polymer materials is detected with CT indices in ranges from 1 to 2,5. Foam samples have sizes of 150x60x (to 30) mm.

Flame spread index, I, is determined for samples with sizes of 320x140x $\delta$ (to 20) mm. The sample is placed at 30° angle to radiant panel, that provides heat flow from 3,3 W/cm$^2$ to 1,25 W/cm$^2$ by length of the sample. Pilot flame (3,1 kJ/s) is used for ignition. The flame spread index is calculaetd as:

$$I=[0,0115\beta\{T_{max}-T_o)/\tau_o\}\{1+0,21\ (1/\tau_i)]^{0.5},$$

where $\beta$ is thermal constant of apparatus; $\tau_o$ and $\tau_i$ are times of flame front reaching the zero and i-th marks; 1 is distance passed by flame front; $T_{max}$ and $T_o$ are maximum and initial temperatures of gases off. Materials are subdivided into 3 groups by flame spread: nonspreading (I=0), slowly - (I up to 20) and fast spreading (I>20) ones.

Smoke formation ability of CPM has been determined by standard method of GOST 24632-81 (corresponds to ISO 5659).

Pyrolythic gas chromatography and mass-spectroscopy have been used for analysis of gaseous products of polymer decomposition in temperature intervals of 300-800°C. Apparatus of GCh-5CO$_2$-2 type have been used also for express-analysis of CPM combustion products[21]. Table 1 demonstrates the examples of fire hazard characteristics of some industrial CPM based on reactionable olygomers.

**Table 1:** Fire hazard indices of some industrial CPM produced in Russia

| CPM Type | $\rho$ kg/m$^3$ | Fire Hazard Indices | | | | | | |
|---|---|---|---|---|---|---|---|---|
| | | LOI % | CT index | I | Smoke form. $D^m{}_{max}$ | Toxicity HCL-50, g/m$^3$ | $T_{ign}$ °C | $T_{si}$ °C |
| Phenolic | | | | | | | | |
| PhRP-1 | 40-120 | 34-40 | 0,6-2,2 | <20 | 14-19$^x$ 4,5-5,5$^{xx}$ | 6,6-14$^x$ 210$^{xx}$ | 470-500 | 540-600 |
| Urea formald. | | | | | | | | |
| MPhP-3 | 15-30 | 30-34 | 1,3 | <20 | 110-250$^x$ | – | 300 | 475 |
| PKZ-30 | 25 | 34 | 1,35 | <20 | 200$^x$ | – | 280-300 | 450-475 |
| Polyurethane | | | | | | | | |
| foam PPU-316M | 50-180 | 20-22,5 | 1,9-4,5 | 29,8 | 520-845$^{xx}$ | 24-29,2$^x$ | 360 | 530 |

x - pyrolysis conditions; xx - combustion ones.
Critical heat flow for PhRP-1 ignition is 30,4kW/m$^2$ (for foam with $\rho$=50 kg/m$^3$).
Effect of CPM porosity on LOI is insignificant.

Phenolic and urea formaldehyde foams are hardly ignitable ones. Phenolic foams of PhRP-1 type are hardly combustible, if their apparent density become more of 80 kg/m$^3$. These foams show very low optical density of smoke. However, high toxicity of combustion products is one of the disadvantage of phenolic foams.

# 4. FLAME SPREAD ALONG THE SURFACE OF CELLULAR POLYMERS

By present time the phenomenon of flame spread along the surface of various combustible materials has been studied to a considerable extent both experimentally and theoretically. Flame spread (FS) is considered as a result of complex interaction of processes of mass and heat transfer and chemical reactions in gas and condensed phases. The rate of combustion wave movement is considered the main FS index.

Fundamental FS investigations were successful for understanding mechanisms controlling the process. Two main mechanisms have been clarified depending on conditions of FS realization - heat transfer from the flame to ignited surface of material and kinetics of gas phase chemical reaction of fuel vapours with oxidizer.

Numerous FS types are usually classified by space orientation of the material, direction of mutual movement of combustion wave and oxidizer flow, combustion hydrodynamics, thermal thickness of sample. One can find reviews of works, devoted to various FS modes in [22-24]. Most of theoretical FS models are thermal, as they describe heat transfer to the surface of combustible material both from its own flame and external source. Heat transfer from flame to the material surface occurs, as a rule, by heat conductivity or convection through gas phase, heat conductivity through a condensed phase and by radiation from flame, either.

To calculate FS rate one should suppose the heat flow, $q''$, to inflammable material surface be expended on limit change of its enthalpy, $\Delta H$, necessary for ignition[25]: $\rho V_{fs} \Delta H \delta = q''l$, where $\rho$ is fuel density; $\delta$ is the thickness of pyrolizable layer; $l$ is surface size affected by heat flow from the flame.

Determination of predominant mode of heat transfer helps to understand main regularities of FS along the material surface. Various external factors, affecting heat transfer from reactional zone to the surface, affect FS rate. External radiation heat, changes in temperature of material and surrounding, heat loss from the article, oxygen concentration in oxidizer flow, the rate of latter, pressure, etc., are among these factors.

The direction of gaseous oxidizer flow in respect to combustion wave influences FS rate as it usually weakens (at opposed flow) or increases (at concurrent one) heat transfer from flame to the surface. Both modes of flame spread can be presented as pyrolysis front spread ($x_p$) and written by the same approximate functional equations. For example, for thermally thick materials we have [26]:

$$V_{fs} = dx_p/d_r = q_f''^2 l_f / (T_i - T_s)^2 \lambda \rho c,$$

where $q''_f$ is heat flow from the flame to the surface at the distance $l_f$ before pyrolysis front; $x_p$ is pyrolysis front coordinate; $T_i$ is ignition temperature; $T_s$ is surface temperature at the $l_f$ distance. At opposed directions of the oxidizer flow and the flame one $q''_f$ and $l_f$ depend largely on thermal properties of the material. FS rate at concurrent oxidizer flow depends to a great extent on heat release rate, which is connected with mass rate of material burning out. At concurrent flow of gaseous oxiidizer and flame movement FS rate increases. The analysis of FS of this mode predicts the power dependence of FS rate on time and pyrolysis zone length in material. CPMs differ from related monolithic materials by their density and thermal physicial properties. The decrease of thermal inertia ($\lambda\rho c$) for CPM should induce the increase of FS rate as it follows from the above mentioned equation.

In this work the characteristics of the downward laminar diffusion FS over PU foam slab are determined by using LOI test apparatus. It was interesting to examine the effect of porous foam structure (apparent density) and also other factors on the FS rate. In order to elucidate the question about the mechanism of heat transfer from the creeping flame to inflammable PU, surface profiles of temperature near the leading edge of the flame are examined in detail. Pt-PtRh thermocouples with noncatalytic coating are used. Micromanipulator positioned thermocouples at 0.25 mm intervals normal to the vertical surface. The special marker is used for the determination of the moment of flame arrival to absolute distance scale. The temperature-time(distance) graphs are recorded by H115 oscillograph. Figure 2 shows temperature distribution profiles obtained for the steady downward flame spread along vertical slab of PU foam having $\rho=49$ kg/m$^3$ at $Y_{O_2}=0,26$.

The foam samples of size of 1.25x1.25x15 cm exhibit the behavior of a thermally thick ones. At the condition studied the thickness of the surface porous layer heated under the flame leading edge is equal to 1.1mm. The surface temperature is 648K. The flame leading edge is disposed above the surface of PU foam at y=1.5mm. Maximum flame temperature is 1220-1370 K.

The temperature data are used to calculate an energy balance for the condensed phase. For laminar flame spread over PU foams samples of small size of heat feedback, $q''_s$, from the flame to inflammable surface is used to raise temperature of the layer, $q''_h$, and to pyrolize the fuel, $q''_s$: $q''_s=q''_h+ q''_{pyr}$

For steady process at $x<0$, $q''_s=q''_h$. The net flux of thermal enthalpy is $q''_f$, where $\delta_{cy}$ is the thickness of porous layer heated; c is heat capacity of foam. In the case considered heat transfer mechanism

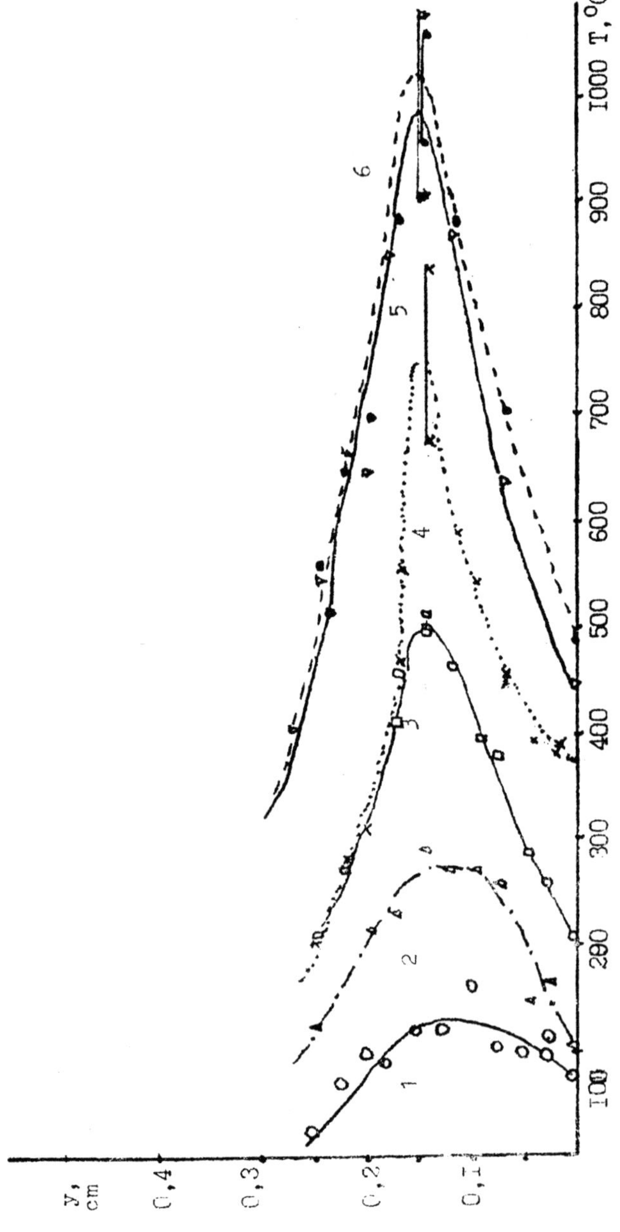


**Figure 2.** Temperature distribution profiles in the gaseous phase at downward flame spread along PU foam sample (r=49 kg/m$^3$). 1.x=-2; 2.x=-1; 3.x=-0.5; 4.x=0; 5.x=+1; 6.X=+1,5 mm.

includes gas and solid phase thermal conduction through cellular surface layer with depth of $\delta_{cy}$. From the experimental data on flame spread rate, temperature distribution profiles and temperature gradients in gas and solid phases the values of $q''_h$, $q''_g$, $q''_{cs}$ are calculated separately as function of x distance.

Figure 3 demonstrates a considerable fraction of heat transfer from flame by the thermal conduction through porous layer near the leading flame edge (to 30-40% of net heat flux from flame). The heat transfer through gas phase is dominant mechanism of flame spread along PU foam surface in this small scale test.

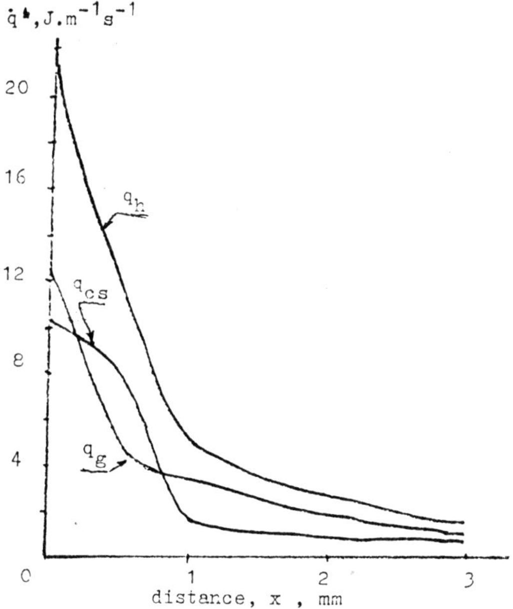


**Figure 3.**  Heat fluxes distribution ahead flame front at downward flame spread along PU foam.

It is found that surface temperature and flame one increase to 675-680K and to 1350-1450 K when oxygen content into oxidizer flow increases to 28%. The thickness of surface cellular layer heated reduces to 0,8 mm. Due to low thermal inertia of PU foam the heat transferred from flame to the inflammable surface of cellular body is accumulated by thin surface layer.

The mechanism of flame spread over charring phenolic foam is characterized by some pecularities. PhRP-1 foam with density of 60

$Kg/m^3$, exposed at 40°C during 10 hours, was studied. The rate of downward flame spread over the foam surface at oxygen content in oxidizer flux of 50% was equal to 0,3 mm/s. The surface temperature under flame leading edge was 740-768K. Maximum surface temperature of carbonized layer (1170K) was realized at a distance of 0.56 mm from flame leading edge coordinate. By using data of temperature profiles into combustion wave, thermal balance was calculated. It is found that heat transfer by the thermal conduction through gas phase is the dominant mechanism of flame spread over charring foam in this small-scale test. The heat transfer from flame by radiation is insignificant (<1% of net heat flux to the surface). Radiation from incandescent char is more important for the heating of surface layer under flame leading edge. However ahead flame edge this mode of heat transfer has no role. For charring cellular polymer we observe the approach of flame leading edge to the surface of foam.

**Table 2.**   The effect of apparent density of rigid foams on FS rate

| | Apparent Density, $kg/m^3$ | | | | | | | |
|---|---|---|---|---|---|---|---|---|
| PU foam | 30.9 | 31.2 | 33.5 | 34.6 | 47.6 | 48 | 48.8 | 74 | 77.8 |
| $V_{fs}$, mm/s ($Y_{ox}$=0.265) | 7.4 | 7.36... | 5,45... | 4.93... | 3.84... | 3.7... | 3.26... | 2.8 | 2.6 |

| | Apparent Density, $kg/m^3$ | | | | | | | |
|---|---|---|---|---|---|---|---|---|
| Phenolic foam ($Y_{ox}$=0.43) | 31.8 | | 49.3 | | 60 | | 81.3 | | 120 |
| $V_{fs}$, mm/s | 1.2 | | 0.70 | | 0.57 | | 0.41 | | 0.26 |

It should be noted that maximum flame temperature at flame quenching limit for all used foams does not exceed 1370K.

In the present work it is found that the flame spread velocity depends on apparent density of foams. In general form the dependence is described by the equation:

$$V_{fs}=K\rho^{-n},$$

where K and n are constants.

Table 2 demonstrates the experimental results, obtained for the flame spread over PU and phenolic foams in the downward vertical directions [27]. The results observed show that chemical nature of cellular polymers as well as their apparent density affects flame spread rate.

The results obtained for rigid PU foams can be described by equation: $V_{fs}=2.8 \times 10^2 \rho^{-1.1}$, mm/s.

For phenolic foams we have the next correlation:

$$V_{fs}=0.636\times10^2\,\rho^{-1.15},\ mm/s.$$

It should be noted that the flame spread rate over foam surface in horizontal direction was practically near to the one in downward vertical one. The value of "n" in above equations almost approach to the theoretical one. Thus cellular polymers of lower apparent density can develop a higher rate of flame spread. A large effect of thermal inertia of cellular polymers on FS rate may be illustrated by the comparison of two rigid PU foams with roughly equal density (49 and 40 $kg/m^3$) and porosity (96.1 and 96.6%) but with different values of thermal inertia ($\lambda\rho c.10^{-3}$ are 2.3 and 1.78, $W^2s/m^4K^2$). FS rate along surface of PU foam with a lower thermal inertia is almost twice as much the value of FS rate for other foam.

The considerable effect on the flame spread rate over cellular polymers is found to have oxygen content into oxidizer flux ($O_2+N_2$). The effect is increased when the apparent density of foam is decreased. As example, the dependence of the FS rate over PU foam having $\rho=49Kg/m^3$ on oxygen content in limits from 25 to 30% vol. is approximated by the equation: $V_{fs}=3.7\times10^5Y^{8.5}_{ox}$, mm/s. For PU foam with $\rho=70\ kg/m^3$ the dependence corresponds to the equation: $V_{fs}=3.4\times10^4Y^{8.5}_{ox}$, mm/s.

# 5. FLAME RETARDANCY OF RIGID FOAMS BASED ON REACTIONABLE OLYGOMERS

The important feature of rigid foams based on a reactionable olygomer is the including of the stages of net polymer formation and cellular macrostructure in united technological cycle. Therefore, one of the general requirements to flame retardeants used for the production of the rigid foams lack negative effects on chemical reactions of network formation and the regimes of foaming. In the opposite case the change for the worse of different exploitational properties of cellular polymers is observed. Let us consider some approaches to flame retardancy of rigid foams studied in the work.

## 5.1. Phenolic Foams

Phenoloformaldehyde foams are widely used as thermal insulation materials in native building industry. At relatively low apparent density (40-80 $kg/m^3$) they are characterized by low values of thermal conductivity, high heat resistance and shape stability. The foams are hardly ignitable materials with low smoke formation. However, phenolic foams have the number of serious shortcomings: low mechanical strength, corrosive activity, unsufficiently high thermooxidative stability, the inclination to smouldering and also high toxicity of combustion and pyrolysis products.

The objects of the present study are phenolic foams of pour out type based on resol phenoloformaldehyde olygomers of cold hardening. For production of flame retardant phenolic foams there are used the components of industrial rigid foams of PhRP-1 type: 1.phenoloformaldehyde resin of PhRV-1A (Technical condition of 6-06-1104-78); 2.foaming and hardening agent of VAG-3 (Technical conditions of 6-05-116-78).

PhRV-1A resin is a mixture of resolic phenoloformaldehyde olygomers with aluminum powder and surface-active substance of OP-10. Foaming and hardening agent of VAG-3 is condensation product of sulphophenylcarbamide with formaldehyde and ortho-phosphoric acid. Typical formulations of the compositions for production of phenolic foams consist of 100 mass.parts of PhRV-1A and 15-25 mass.parts of VAG-3. Foaming is the result of hydrogen formation at the interaction acid with aluminum powder.

The purpose of the present work was to obtain the flame retardant phenolic foams with improved performances. The next approaches have been realized:

1.  The using of reactionable phosphorus-containing substances, which permits to obtain polymer matrix with interpermeated or semi-interpermeated network structure (network-network or network-linear polymer).
2.  The using of the additives with multifunctional mechanism of the action, which permit to affect the different properties of initial composition and foam.

Bis(2-chloroethyl) 1-bromovynilphosphonate (bromophos-1), bis (2-hydroxyethyl) methacrylphosphonate (PhEM), the product of the interaction of pentaerythritol and alkylphosphonic acid (phosphopolyol) as reactionable flame retardants have been investigated. The substances are well compatible with PhRV-1 resin. They do not practically affect both foaming and hardening regimes. Foams obtained by the way have homogeous, small cellular structure. Table 3 shows the properties of phenolic foams modified. The results for the modification of PhRP-1 foam by poly (1,4-phenylenephenyl-phosphonate (PPhPh) are also given. PPhPh (Rhone-Poulenc product) is linear polymer with $M_n$=10000-11500. PPhPh thermal decomposition is observed above 375°C (Figure 4). The additive behaves as disperse filler in resin, therefore the small increase of foam apparent density with PPhPh content takes place.

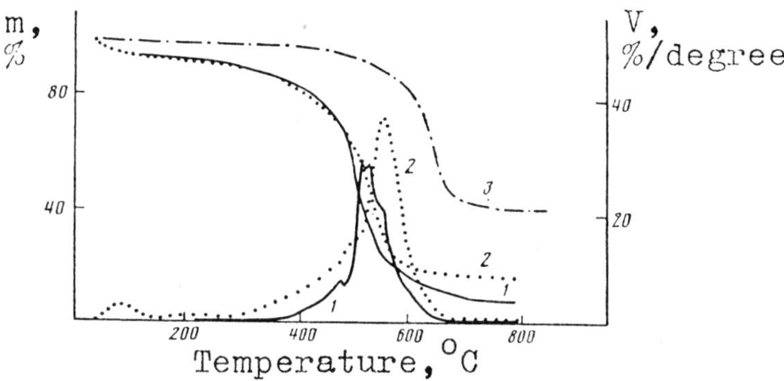

**Figure 4.** TG and DTG curves of decomposition for phenolic foam (1), phenolic foam with 10% PPhPh (2) and PPhPh (3).

Phenolic foams modified are hardly combustible, no smouldering after removal of flame. When phosphorus content in foams run up 0,8-1% mass, critical thermal flux of ignition increase to 45-50 kW/m$^2$. Mechanical strength of foams modified is higher than one of ordinary foams. Air temperature, above which study smouldering for foams ignited by pilot flame appears, is also higher (310-320°C instead of 250-270°C for ordinary foams).

**Table 3.** The properties of flame retardant phenolic foams[28-31]

| Indices | Phosphorus containing flame retardant | | | |
|---|---|---|---|---|
| | PhEM | Bromophos-1 | Phosphopolyol | PPhPh |
| FR content, pph by weight | 3,3-16,7 | 2,4-14,8 | 7,0-10,5 | 1-10 |
| Apparent density, kg/m$^3$ | 65-56 | 45-65 | 58-114 | 60-85 |
| Compressive strength, kPa | 150-230 | 100-230 | 110-310 | 115-230 |
| Water absorption during 24h, % | 12-16 | 14-20 | 12-18 | 14-29 |
| Heat resistance, °C | 170-175 | 160-165 | 170-175 | 165-170 |
| Acid number, mg KOH/g | – | – | 2,8-8,7 | – |
| Decomposition temperature, °C | 280-300 | 265-275 | 280-300 | 280-300 |
| Mass losses, % (Fire tube test) | 12,8-14,2 | 10,0-15,4 | 11,4-16,0 | 10-16 |
| CT index | 0,2-0,47 | 0,17-0,37 | 0,15-0,21 | – |
| LOI, % vol. | 41,7-45,0 | 35,4-49,5 | 48,5-58 | 42,4-48,3 |
| Smouldering temperature, °C | – | – | 300-320 | 300-310 |

**Table 4.**   The PPhPh effect on the composition of thermal degradation products of phenolic foam[31]

| Product | Content, % weight, at temperature, °C | | | | |
|---|---|---|---|---|---|
| | 300° | 400° | 500° | 600° | 700° |
| CO | 0,07/0,03 | 0,08/0,08 | 2,7/1,3 | 3,9/4,6 | 4,0/3,8 |
| $CO_2$ | 0,1/0,07 | 0,29/0,13 | 3,1/2,8 | 4,2/3,1 | 3,5/2,2 |
| $CH_4$ | 0,2/0,14 | 0,25/0,27 | 4,0/2,3 | 5,2/4,5 | 5,9/6,7 |
| $H_2O$ | 0,87/0,2 | 0,91/0,66 | 4,1/4,0 | 5,9/7,5 | 6,6/7,5 |
| $CH_2O$ | 0,06/0,04 | 0,44/0,25 | 1,0/1,24 | 1,1/1,9 | 0,8/1,7 |
| benzene | 1,0/0,6 | 2,0/1,6 | 5,3/9,8 | 9,9/12,1 | 17/21,4 |
| toluene | -/trace | -/trace | trace/2,4 | trace/3,1 | -/0,5 |
| phenol | 4,6/3,8 | 16,9/11,2 | 34,6/29,7 | 45,9/48,2 | 48/47,1 |
| o,m,p-cresols | -/trace | -/trace | trace/2,4 | trace/3,1 | -/0,5 |
| fraction with $T_{vol}>200°$ | 92,3/95,1 | 78,8/85,8 | 47,0/42 | 21,9/11,2 | 9,4/7,4 |
| nonvolatile residue | 87,2 | 82,5 | 56,9 | 37,5 | 36,8 |

numerator - without, denominator - with 10% PPhPh; 2 min. isothermal heating in inert atmosphere

The comprehensive study of thermal decomposition of phenolic foam modified by PPhPh is carried out[31]. It is determined that PPhPh affects the content of volatile products and char yield (Table 4). The balance on amounts of volatiles confirms that all phosphorus of PPhPh at high temperature pyrolysis of phenolic foam remains in condensed phase. Flame retardancy mechanism of phenolic foams by PPhPh is connected with the increasing of char yield and with the formation of surface polyphosphate layer at pyrolysis of cellular polymer, the retardation of thermooxidation of carbonized residue.

Industrial resins, used for phenolic foam production, contain the light volatile toxic compounds such as free phenol (to 9%) and formaldehyde (to 3,5%). During foaming and hardening of foam composition the high temperature (to 110°C) is developed as a result of exothermal reactions. The toxic compounds volatilize and make worse the labour conditions. Short period for a keeping of phenolic resins is one of their demerits.

Authors showed that the application of small amounts (0,02-2pph) of metal halides or metal borohalides as such as $AlF_3$, $SnCl_2.2H_2O$), $Zn(BF_4)_2.6H_2O$ permits to decrease the apparent density of foams, the content of free phenol in the articles and also to use unconditional resins after long keeping[32,33]. The formation of metal salt complexes with phenol has been shown by UV-spectroscopic analysis of model systems including phenol.

Complexing activity of salts is decreased in the next order:

$$SnCl_2 > Zn(BF_4)_2 \geq AlF_4 > CuF_2 > MnCl_2 \approx Ni(BF_4)_2 > NiCl_2$$

At introduction of 0,02-0,05 pph $AlF_3$ the apparent density of foams obtained in comparison with foam without additive decreases from 51 to 45,3-38 $kg/m^3$, however, LOI increases from 37,7 to 40,3-41,5%. Compression strength of foams remained on the level of 135-140 kPa. The considerable decrease of free phenol content in foam is observed[32].

It is found that the salt's additives permit to obtain phenolic foams with good performance on basis of resins aging above 6 months. As example, for ordinary foam formulation, but with the application of PhRV-1 resin exposed at 20-25°C for 6 months, foam with apparent density of 165 $kg/m^3$ is produced. The use of $ALF_3$ additive (2,5pph) permits not only to develop phenolic foam with $\rho=75$ $kg/m^3$ but also to decrease free phenol content from 5 to 1,1%[32].

The effects are connected with the influence of the additives in reological properties of polymeric composition, on the reactions of hydrogen evolution and polycondensation ones of resin components. Phenolic foams modified by the way have structure with some content of closed cells.

Flame retardancy of rigid phenolic foams can be improved by the using of high temperature stabilizers against thermooxidative decomposition of cellular polymer.

It is shown that cupric complex of macroheterocyclic compound is an effective stabilizer for thermooxidative decomposition of phenolic foams at temperature above 200°C. The additive of 0,25-5,0 pph permits to decrease the ignitability of phenolic foams at high temperatures of air environment[34].

## 5.2. Urea Formaldehyde Foams

Urea formaldehyde foams (UFF) have more narrow interval for working temperatures (from -20° to + 130°C) than phenolic foams. The small mechanical strength, high water and moisture absorption, considerable shrinkage, the release of free formaldehyde at the production and the article's exploitation are the main shortcomings of UFF. Chemical modification of UFF and the using of inorganic phosphorus and nitrogen-containing flame retardants are a general approach to UFF flame retardancy. Hardly combustible UFF's with closed cellular structure are developed by chemical modification and in addition by the using of 1-2 pph of phosphorus containing substances[35]. At flame retardant content above 2-2.5 pph the strong increase of times for the beinning and the termination of foaming process is observed.

UFF's developed ($\rho$=60-120 kg/m$^3$) have next flammability indices: LOI is 40-41%; ignition and self-ignition temperature are 285-305° and 465-475°C correspondingly; CT index is 0,31-0,55. It was turned a great attention to working out of the ways for the decrease of formaldehyde off gassing during the UFF production.

We suggested to affect the strengthening of intermolecular interaction for formaldehyde to maintain into associated state and then gradual inclusion of it into network formation process. The preliminary positive results are obtained by the way.

## 5.3. Rigid Polyurethane Foams

Rigid polyurethane foams (PUF) are occupying the leading position among other cellular polymers used in building and transport as thermal insulation materials. These cellular polymers are popular due to simple technology of their production, relatively low cost and high insulation properties.

Reactive and additive flame retardants, containing halogen, phosphorus and nitrogen, have been used extensively in such foams to decrease their flammability. In recent years there has been a growing interest in the development of rigid PUF which perform better in fire tests and conditions. This has led to a general tendency to limit the use of halogen-based flame retardants in spite of their relatively large effectiveness in improving the reaction to fire of polymer materials. It is known that fire behaviour of single components does not characterize materials combined to composite building elements. However, fire retardancy of each of the components permits to improve fire safety performance of composites on the whole.

In the present work the effect of halogen free reactive flame retardant on a main technological, exploitational properties and fire safety indices of rigid PUF has been investigated[36,37].

A starting formulation for producing a rigid PUF modified includes the next components:

|  | Part by weight |
|---|---|
| polyisocyanate (Technic.condit. TC 6-03-296-78) | 125 |
| polyol Laprol 503 or Laprol 805 (TC 6-05-1679-74) | |
| N-containing polyol Lapromol-294 (TC 6-05-1681-74) | 100 |
| dimethylethanolamine | 2 |
| freon-11 (TC 6-02-727-78) | to 30 |

Phosphorus-containing flame retardant is used in limits of 10-70 parts. Phosphorus-containing olygoester ethoxylated (Phospolyol) has viscosity of 18,43 cPa.s at 75°C. It contains 15,2% P, 0.08% water and 11,0% hydroxyl groups. It is found that phospolyol decreases a start foaming time from 28 to 21s. and termination time of foaming from 110 to 70s (for 70 parts of phospolyol).

Figure 5 demonstrates that flammability indices of PUF modified depend on phosphorous contents. Optimum P contents in foam is about 2,5%. In this case PUF has the next properties:

| | |
|---|---|
| the apparent density, kg/m$^3$ | 50-55 |
| compressive strength, kPa | 340-360 |
| water absorption for 24 hours, %vol. | 1,7-1,8 |
| thermal conductivity, W/m.K | 0,028 |
| Limiting oxygen index, % vol. | 27,6-27,8 |
| CT index | 1,34-1,4 |
| smoke optical density D$^m_{max}$ at combustion | 237-245 |
| at pyrolysis | 103-107 |

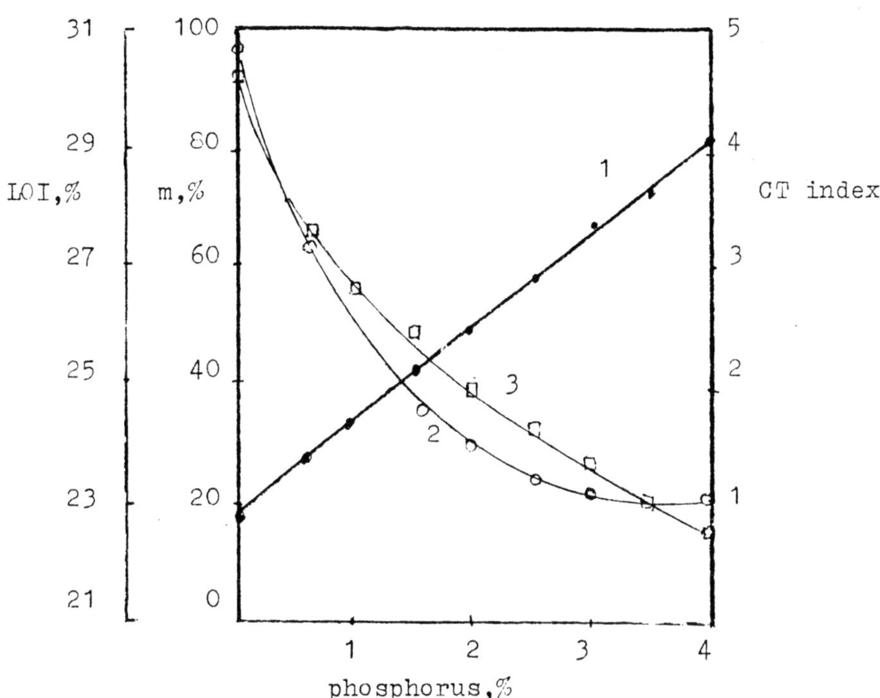


**Figure 5.** The dependence of flammability indices for PU foams on phosphorus contents. 1 - LOI; 2 - mass losses at fire tube test; 3 - CT index.

The composition of pyrolysis products of FR foam does not practically change with increase of P contents (Table %):

**Table 5.**    The composition of pyrolysis products of rigid FR foam (2,5%P)

| mass loss at 400°C % | Contents, % | | | | | | | |
|---|---|---|---|---|---|---|---|---|
| | CO | $CO_2$ | $H_2O$ | $CH_4$ | $C_2H_6$ | $C_2H_4$ | $C_3H_6$ | HCN |
| 83,8 | 0,37 | 1,8 | 0,61 | 0,11 | 0,06 | trace | 0,22 | – |
| | 2,44 | 4,69 | 0,97 | 1,40 | 0,17 | 0,02 | 0,31 | 0,028 |

numerator - at 400°, denumerator - at 700°C

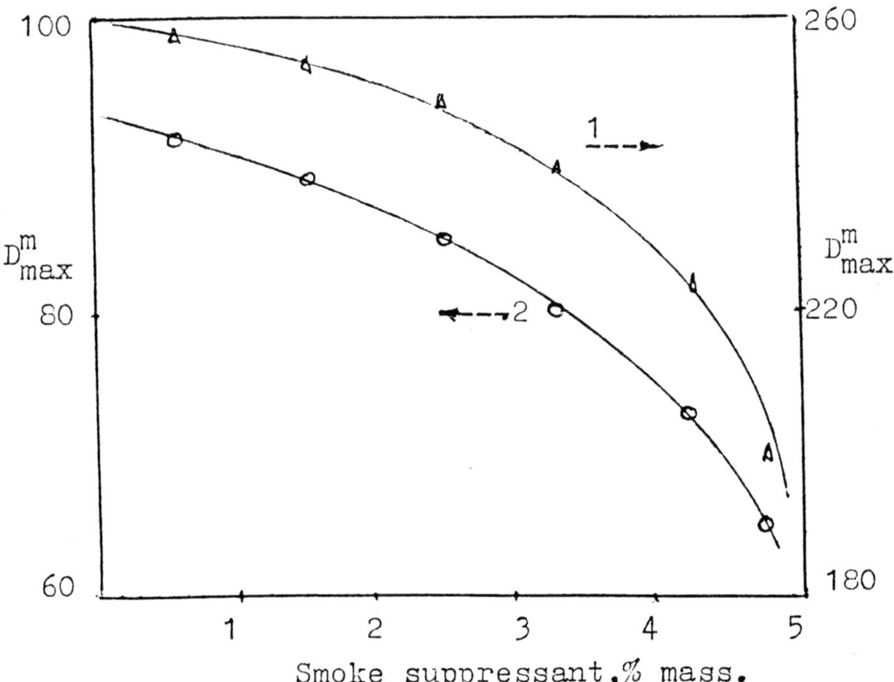


**Figure 6.**   The effect of natrium trimolibdate on smoke formation at combustion (1) and pyrolysis (2) of phosphorus containing PU foam.

Thus, PUF modified are hardly ignitable materials. Toxicity of high temperature pyrolysis products for PUF is connected with increasing HCN content. To decrease the foam ability to smoke formation it was supposed to introduce additionally three molybdate of natrium (Figure 6). This compound does not affect LOI of foam, but decreases toxicity of pyrolysis and combustion products. As example, for 2% of the smoke depressant CO contents decrease from 0,37 to 0,06% vol. Critical heat flow for the PU foam ignition increases from 18,1 (without FR) to 30,5 kW/m$^2$.

# 6. CONCLUSION

Some aspects of combustion behavior and flame retardancy of rigid cellular polymers based on reactionable polyfunctional olygomers have been considered. The approaches, proposed by us to solve the flame retardancy problem for cellular polymers, permit to develop rigid phenolic foams, urea formaldehyde and PU foams with high performance.

# REFERENCES


1. *Saechtling international plastics handbook for technologist, engineer and user*, 2nd Ed., Hanser Verlag, Munchen, 1987.
2. **V.N. Felichkina**, *Chemical industry abroad*, 1987, N9, 41 (in Russian).
3. *Palontorjunta*, 1982, vol. 33, 560.
4. *Brandvaern*, 1987, vol. 13, 17.
5. **A.A. Berlin, F.A. Shutov**. *Chemistry and technology of foamed polymers*, Science, Moscow, 1980 (in Russian).
6. **A.A. Berlin, F.A. Shutov**. *Foamed polymers based on reactionable olygomers*, Chemistry, Moscow, 1978 (in Russian).
7. **A.A. Berlin, F.A. Shutov**. *Structural Foamed Plastics*, Chemistry, Moscow, 1980 (in Russian).
8. **A. Schauer et al**. *Stavebnicky casopis* 1967, vol. 15, 245.
9. **R.H. Harding**. *Resinography of cellular plastics*, ASTM Spec. tachn. publ. N414, ASTM, Philadelphia, 1967, 3.
10. **Yu.M. Tovmasyan, F.A. Shutov et al**. Reports of USSR Academy of Sciences, 1982, vol. 263, 156 (in Russian).
11. **A.A. Berlin, I.I. Chaikin, F.A. Shutov**. *Plastmassy*, 1982, N2, 14.
12. **A. Gunningham, D.J. Sparrow**. *Cellular polymers*. 1986, vol. 5, N6, 327.
13. **R.E. Scochdopole**. *Chem. eng. Progr.* 1961, vol. 57, N10, 55.
14. **M.E. Aerov, O.M. Todes, D.A. Norchinsky**. *Apparatus with stationary grained layer*, Chemistry, Leningrad, 1979 (in Russian).
15. **R.Williams, C.M. Aldao**. *Polym. Eng. Sci.* 1983, vol. 23, N6, 293.
16. **M.A. Schuetz, L.R. Glicksman**. *J. Cell. Plast.*, 1984, vol. 20, N2, 114.

17. **A. Cunningham, G.M.F.** Jeffs et al. *Cellular Polym.* 1988, vol. 7, N1, 1.
18. **A.D. Golikov, V.M. Ivanov.** *Chemistry and Technology of production processing and application of polyurethanes.* Chemistry, Vladimir, 1984, 86 (in Russian).
19. **A.G. Dementyev, O.G. Tarakanov, M.I. Fedotova**, ibid, p. 43.
20. **C.J. Hilado.** *Flammablity Test Method Handbook,* Tachnom. Publ. 1973.
21. **GOST** 12.1.044-89 *Fire hazard of substances and materials.*
22. **I.S. Wichman.** *Progr. Eng. Comb. Sci.* 1992, vol. 18, 553.
23. *Combustion Sci. Technology,* 183, vol. 32, N1-4, pp. 1-209.
24. **A.C. Fernandez-Pello.** *Comb. Sci. Techn.* 1984, vol. 39, 119.
25. **F.A. Williams.** 16th Symposium (Intern.) on Combustion, Pittsburg, Comb. Inst., 1977, 1281.
26. **J. Quintiere.** *Fire and Materials,* 1981, vol. 5, N2,52; 1985, vol. 9, N2, 65.
27. **R.M. Aseeva, L.V. Rubun, L. Zabski.** *Intern. J. Pol. Mater.* 1990, vol. 14, 127.
28. **R.M. Aseeva, V.A. Ushkov, L.V. Ruban, G.E. Zaikov, et al.** SU author. sert. N825557, Int. $Cl^3CO8$ J 9/06, SU Bulletin of Inventions 1981, N16.
29. **R.M. Aseeva, V.A. Ushkov, L.V. Ruban, G.E. Zaikov, et al.** SU author. sert. N664436 SU Bul. Inv. 1979, N19.
30. **V.A. Ushkov, R.M. Aseeva, R. Andrianov, G.E. Zaikov. et al.** SU auth. sert. N1407021, 1988.
31. **R.M. Aseeva, M.I. Artsis, G. Vivan, G.E. Zaikov.** *Vysokomol. soed.* 1987, vol. 29A, N12, 2596.
32. **R.M. Aseeva, V.A. Ushkov, G.E. Zaicov et al.** SU author sert. N872532, *SU Bull. Inv.* 1981, N38.
33. **M.G. Bruyako, R.M. Aseeva, V.A. Ushkov et al.** SU author sert. N784303, *SU Bull. Inv.* 1980, N44.
34. **F.A. Shutov, R.M. Aseeva, V.V. Ivanov et al** SU author sert. N535323 *SU Bull. Inv.* 1976, N42.
35. **V.A. Ushkov, O.V. Zacharova, V.V. Samoshin et al.** *Plastmassy,* 1989, N7, 72.
36. **V.A. Ushkov, R.M. Aseeva, V.I. Kalinin et al.** *Plastmassy,* 1984, N9, 21.
37. **V.A. Ushkov, R.M. Aseeva, V.N. Vorobyev et al.** *Plastmassy,* 1989, N7, 45.

# THE EFFECT OF ADDITIVES OF METALS AND THEIR DERIVATIONS UPON POLYMER THERMODEGRADATIONS


**L.V. Ruban and G.E. Zaikov**
*Institute of Chemical Physics, Russian Academy of Sciences,*
*4 Kosygin Street, Moscow 117977*


## INTRODUCTION

Polymer pyrolysis accompanied by forming of fuel gases which are generally dangerous during fire. The chemical elements most widely associated with an ability to inhibit the combustion of polymers are non-metals, principally chlorine, bromine, phosphorus, boron and sulphur [1,2], which are noxious toxic materials.

The studying of effect of additives of metals and their derivations upon polymer thermodegradations closely are connecting with the search of ecological pure retardants for polymers [3,4].

## Metals

The effect of metals Au, Cu, Ga, In, Ge, Sn, Sb, Pd and Bi, dispersing in metylmetacrilat (MMA), which was polymerizing in presence azoisobutyronitril (AIBN), have been studied in the work [5]. Colloid synthesis using a metal atom reactor authors reported in the works [6-8].

Authors supposed following reactions:

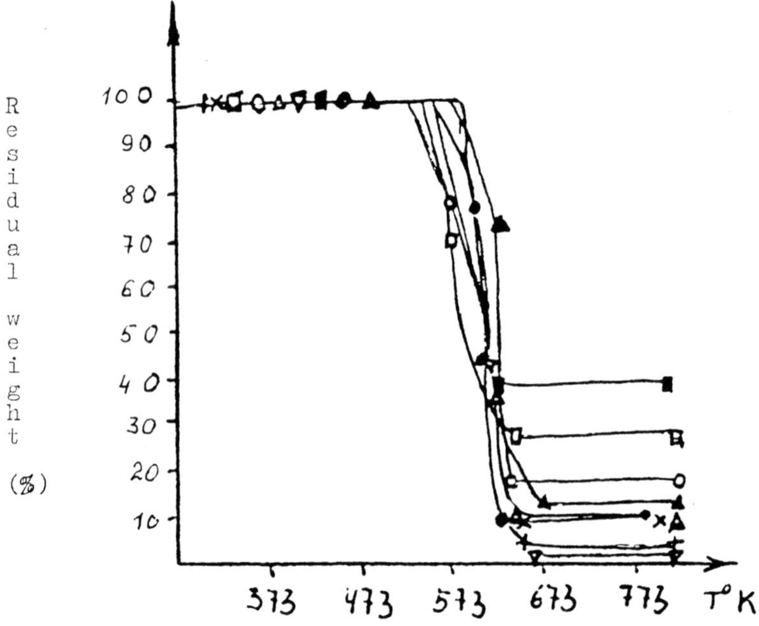

There are only 2% metals in polymer samples. Polymer molecular mass are varying from $2.7*10^5$ to $5*10^5$.

TG-curves of this polymer sample are illustrated in Figure 1. Table 1 shows the thermal decomposition temperatures ($T_D$) for each polymer.

**Figure 1.** Thermogravimetric curves of metal poly (methyl methacrylates) at a heating rate of $10°C/min^{-1}$: ▲, Bi-PMMA; ■, Sn-PMMA; +, Au-PMMA; x, Pd-PMMA; □, Cu-PMMA; o, Ge-PMMA; Δ, Ga-PMMA; ∇, In-PMMA; •, Sb-PMMA.

**Table 1.** Effective kinetic parameters of thermal degradation metal-polymethylmetacrilates.

| polymer | Z $(s^{-1})$ | Ea $(kJ.mol.^{-1})$ | n | $T_D$ $(K)$ |
|---|---|---|---|---|
| Au-PMMA | $2.6*10^4$ | 99.33 | 0 | 573 |
| Bi-PMMA | $2.3*10^2$ | 78.91 | 0 | 598 |
| In-PMMA | $1.4*10^2$ | 74.27 | 0 | 548 |
| Cu-PMMA | $2.4*10^2$ | 70.67 | 0 | 513 |
| Ge-PMMA | $4.8*10^1$ | 64.10 | 0 | 498 |
| Sb-PMMA | $2.5*10^1$ | 63.97 | 0 | 548 |
| Pd-PMMA | $1.5*10^1$ | 61.50 | 0 | 523 |
| Sn-PMMA | $1.4*10^1$ | 61.09 | 0 | 523 |
| Ga-PMMA | 4.3 | 51.58 | 0 | 473 |

It is interesting to note that Au-PMMA and Bi-PMMA present the highest $T_D$ values (573 and 598 K). The metalpolymer degradation in inertial atmosphere authors described as reaction of the zero law.

Preexponential factors (Z) and effective activation energies ($E_a$) for metalpolymer degradations shown in the Table 1. The influence of metal character on the effective activation energy of the degradation process was demonstrated by this Table. Au-PMMA is very stable system as Au is stable metal to the oxidation. Ga opposite is lightly oxidizing metal, and there are the lowest effective activation energy and thermal decomposition temperature of Ga-PMMA.

## METAL DERIVATIONS

Wilkie and Mittleman for many years are studying the effect of additives of metal derivations upon polymer thermodegradations for possible using this compounds as retardants of polymer combustion. Particularly they have studied blends with compounds of Rh [9,10] and Co [11].

Let's take into consideration some papers which have been published by this author in last years.

## TRANSITION-METAL HALIDES

In paper [12] authors reported the investigation of reaction between PMMA and a metal in a highest oxidation state, chromium (III) chloride, both in the anhydrous and hydrated form.

The reactants were added to the tube and the tube was then thoroughly evacuated for at least 2 h on a high-vacuum line. The tube was sealed off from vacuum line, placed in a muffle furnace, and heated for 2 h at the desired temperature. At the end of the heating period, the tube was carefully removed from the oven and placed in liquid nitrogen.

Pressure-volume-temperature measurements and IR spectra of the noncondensable gaseous products indicate that approximately 6% of the charge has been converted to methane and carbon monoxide. The condensable gases represent about 44% of the starting charge and consist of monomer, $CO_2$, HCl and water.

The chloroform-soluble fraction is a brown viscous material that is about 13% of the mixture. Sodium fusion indicates the presence of chloride, whereas a spot test for chromium shows its absence. The IR spectrum indicates the presence of carbonyl and carboxylate moieties.

The chloroform insoluble fraction was 36% of the starting materials. Its IR spectrum shows the presence of anhydrides with absorptions at 1800, 1760 and 1150 $cm^{-1}$.

An X-ray powder pattern shows the presence of CrOCl and $CrO_3$.

Hydrous chromium chloride, $CrCl_3 {}^*6H_2O$, exists as a variety of coordination isomers: $[Cr(H_2O)_6]Cl_3$, $[Cr(H_2O)_5Cl]Cl_2 {}^*H_2O$, and $[Cr(H_2O)_4]Cl_2 {}^*H_2O$. Each of these shows slightly different behavior upon thermolysis [13]. Below 100°C, the first and third isomers lose only waters, while the second isomer loses only HCl. Since the chromium chloride used in this study loses both HCl and $H_2O$ below 100°C, it is apparently a mixture of hydration isomers. Between 200 and 300°C, each of these isomers will produce chromium hydroxy dichloride, $Cr(OH)Cl_2$. At higher temperatures, chromium oxychloride, CrOCl, is obtained. At elevated temperatures, anhydrous chromium chloride can lose a chlorine atom and produce chromium (II) chloride, $CrCl_2$ [14]. Thus, there may be several species present in this reaction system and it is difficult to identify those actually involved.

Approximately 13% of the initial weight of the hydrous blend was found to be chloroform soluble. When PMMA is pyrolyzed alone in sealed tubes, substantial amounts of oligomers are produced. When PMMA is mixed with $CrCl_3$, simple oligomeric products are not found.

The TGA trace for a 1:1 mixture by mass of $PMMA/CrCl_3 {}^*6H_2O$ is shown in Figure 2. A relatively gradual weight loss occurs until a temperature of about 250°C; above this temperature, gradation occurs more rapidly. Approximately 22% of the sample is lost up to 250°C. Acetone, water, HCl and monomer evolution are detected. Loss of acetone is simply residual solvent from the preparation of the blend. Water and HCl results from thermolysis of the hydrous $CrCl_3$. When anhydrous $CrCl_3$ is used, water is absent but HCl is still observed. Monomer results from the depolymerization of PMMA. Up to 250°C degradation of the individual components occurs with no apparent interaction between them.

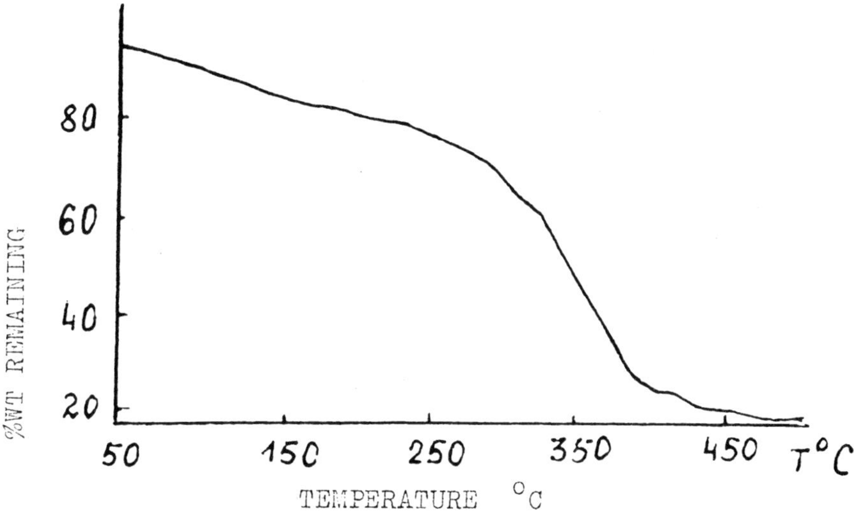


**Figure 2.** TGA curve for a 1:1 by mass blend of hydrous $CrCl_3$/PMMA. The heating rate is 20°C/min.

An additional 62% weight loss occurs between 250 and 500°C. The evolved gases consist of monomer, carbon dioxide, carbon monoxide, an unidentified organic acid, and a trace of methane. Monomer is the principal component evolved until about 430°C. Above this temperature quantity of monomer become less intense and bands in IR spectrum due to $CO_2$, CO, and an aliphatic acid are more intense. For the hydrous chromium compound, HCl evolution is evident at 150°C, from degradation of the hydrous $CrCl_3$, and again above 450°C. With the anhydrous $CrCl_3$, HCl evolution commences at 430°C and it continues until the highest temperature used in this study: 500°C.

The nonvolatile residue that remains at 500°C represents 16% of the starting mass. If the product were $CrCl_2$, then the residual mass would be 23%; COCl would leave 20% residue. A recovery of only 16% indicates that none of these are, themselves, final reaction products.

Analyzing received results, authors explain in the following way. Scheme 1 delineates a pathway for the initial interaction of chromium chloride with PMMA.

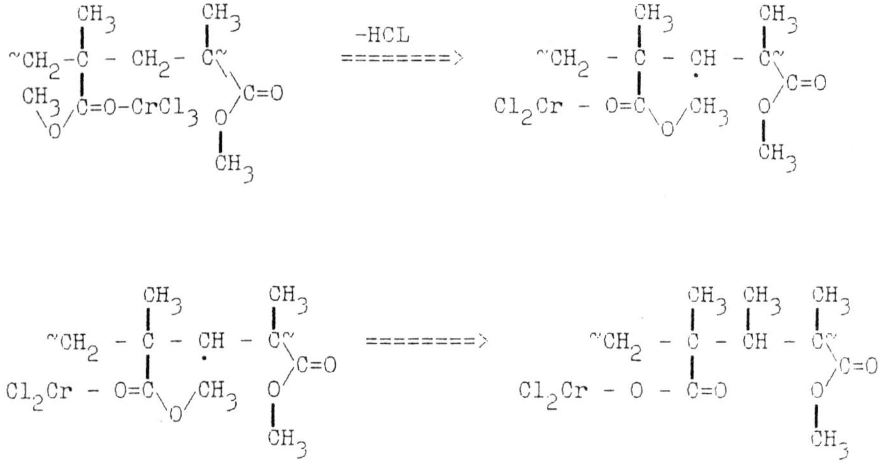

Scheme 1

The first step is probably coordination of $CrCl_3$ to the carbonyl of the PMMA. The work of McNeill and McGuiness on $ZnBr_2$ [15-16] and the work of Wilkie and Mittleman [17] have both suggested an initial coordination to the carbonyl. It is known that $CrCl_3$ can thermally lose a chlorine atom [14]. The liberated chlorine atom may remove a hydrogen atom from the main chain of the PMMA, leaving a radical site on the PMMA chain. The presence of HCl suggests that this reaction occurs rather easily.

The formation of CO, $CO_2$, and $CH_4$ from PMMA degradation have been explained by Manring [18-21] as arising from the cleavage of the carbomethoxy from the main chain of the polymer generating a unstable carbomethoxy radical that can decompose to generate CO and methoxy radical or $CO_2$ and a methyl radical. This is illustrated in Scheme 2.

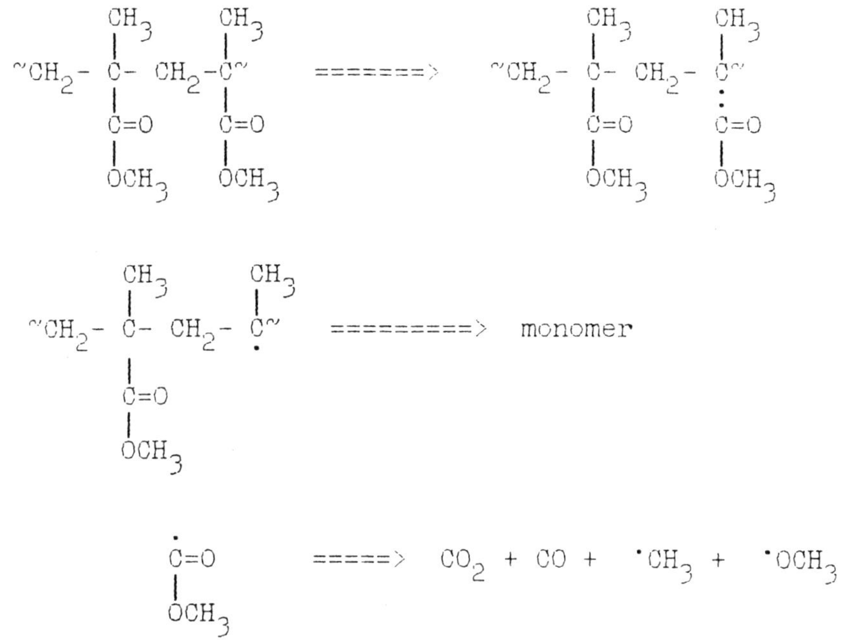

$$\sim CH_2 - \underset{\underset{\underset{OCH_3}{|}}{\underset{C=O}{|}}}{\overset{\overset{CH_3}{|}}{C}} - CH_2 - \underset{\underset{\underset{OCH_3}{|}}{\underset{C=O}{|}}}{\overset{\overset{CH_3}{|}}{C}}\sim \;=======>\; \sim CH_2 - \underset{\underset{\underset{OCH_3}{|}}{\underset{C=O}{|}}}{\overset{\overset{CH_3}{|}}{C}} - CH_2 - \underset{\underset{\underset{OCH_3}{|}}{\underset{C=O}{|}}}{\overset{\overset{CH_3}{|}}{\overset{C}{\cdot}}}\sim$$

$$\sim CH_2 - \underset{\underset{\underset{OCH_3}{|}}{\underset{C=O}{|}}}{\overset{\overset{CH_3}{|}}{C}} - CH_2 - \overset{\overset{CH_3}{|}}{\underset{\cdot}{C}}\sim \;=========>\; \text{monomer}$$

$$\underset{\underset{OCH_3}{|}}{\overset{\cdot}{C}=O} \;=====>\; CO_2 + CO + {}^{\cdot}CH_3 + {}^{\cdot}OCH_3$$

Scheme 2

The formation of olefins has been noted in this reaction by carbon
NMR spectroscopy. The origin of these olefinic bands is believed to be
due to the degradation of the polymer to produce unsaturated oligomers
of PMMA.

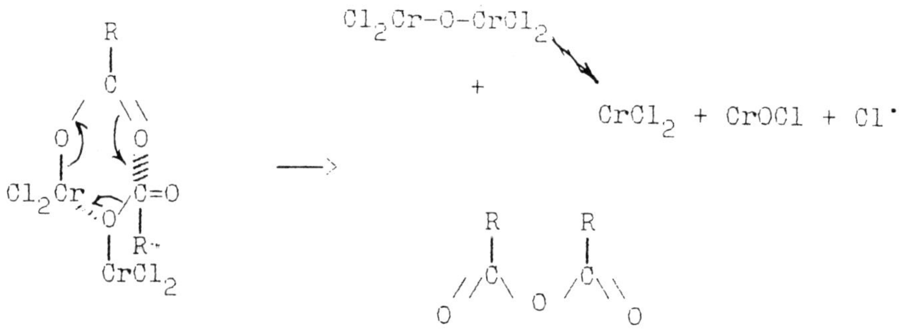

$$Cl_2Cr-O-CrCl_2 + \;\longrightarrow\; CrCl_2 + CrOCl + Cl^{\cdot}$$

Scheme 3

All volatile products have been accounted for by the above mechanism; however, the formation of anhydride and chromium oxides are as yet not explained. As noted above, the pyrolysis of a chromium carboxylate salt generates anhydrides. The interaction of two chromium carboxylate salts may produce anhydride with the elimination of chromium oxides. A possible scheme what in this reaction may occur is shown in Scheme 3.

This process involves the interaction of two chromium carboxylate salts with the formation of anhydrides, $CrOCl$, $CrO_3$, and a chlorine atom. The generation of the chlorine atom implies that another hydrogen atom may be abstracted from the PMMA chain, generating main-chain radicals that may degrade or cross-link. It is of interest to note that anhydride groups are detected even at 270°C. Since this is the last step in the mechanistic pathway, this implies that all of the steps must occur almost simultaneously. Following initial hydrogen abstraction by a chlorine atom, the subsequent steps of methyl migration, reaction with a chromium salt, and the degradation of that salt must immediately occur.

So, a variety of reactions occurs when a blend of chromium chloride and PMMA is pyrolyzed. The most important reactions seem to be (1) the coordination of chromium chloride to the carbonyl of PMMA and subsequent loss of a chlorine atom; (2) abstraction of a hydrogen atom from PMMA by this chlorine atom; (3) a concerted migration of the methyl ester to the main chain of the PMMA, leading to the formation of a chromium carboxylate salt; (4) degradation of the chromium carboxylate salt to give anhydrides; and (5) stabilization of the polymer from the chromium carboxylate formed in the previous step or stabilization by cross-linking of main-chain radicals formed in the previous step. The interaction of chromium chloride and PMMA does not offer any significant reduction in the formation of monomer and, thus, it is unlikely that it can effectively function as a flame retardant.

In the work [22] the thermal degradation of PMMA, in the presence of manganese chloride has been studied by sealed tube reactions and thermogravimetric analysis. A 1.0-g sample of PMMA was introduced into a break seal vessel and heated to 400°C for 2 h. The noncondensable gas was 0.5 mmol CO with a trace of $CH_4$. Condensable gases consisted of $CO_2$ and monomeric methyl methacrylate.

A 1.0-g sample of PMMA and 1.0 g anhydrous $MnCl_2$ were combined in a standard vessel and heated to 400°C. The vessel was opened on the high vacuum line and 0.5 mmol gas was found. Infrared spectroscopy showed the presence of CO and traces of methane. The condensable gas consisted of carbon dioxide and monomeric methyl methacrylate with traces of methyl chloride and methanol. All gases were weighed together, 0.67 g, then the more volatile gases were allowed to escape and 0.30 g monomer were found. The chloroform soluble fraction had a mass of 0.18 g. The infrared spectra [23] showed the presence of ester and anhydride. The insoluble fraction has a mass of 0.85 g, which was slurried with methanol to dissolve the excess $MnCl_2$. The resulting material showed the presence of $MnO_2$ by powder pattern and IR showed an organic salt.

The degree of mixing between the manganese chloride and the polymer is of great importance in monomer formation [24]. When bulk $MnCl_2$ and PMMA are simply poured together and heated under vacuum to 400°C, the monomer yield is less (30-35%) than that observed for PMMA alone (50-70%). It should be noted that these sealed tube reactions do not yield 100% monomer as normally observed for PMMA degradation, probably due to the oligomerization of the monomer. More intimate mixing of the reactants leads to decreased monomer yield. When the two are ground in a mortar and pestle, monomer yield is reduced; when the two components are mixed in solution and then evaporated to give solid, the monomer yield is essentially nil.

The molecular weight of the chloroform-soluble fraction was investigated by viscosity measurements [25] and was found to be about 10% that of the starting polymer. One must construe from this that a significant fraction of the polymer does react to form monomer and that this is then repolymerized to yield relatively low-molecular weight oligomer. Presumable, $MnCl_2$ catalyzes this reoligomerization. In TGA experiments, will be shown that monomer is formed and is able to immediately escape from the system. In sealed tube reactions, product monomer may not escape and is available for further reaction. The intimate mixing that may be achieved by solvent mixing of the polymer and additive must produce a relatively efficient catalyst for reoligomerization.

The combination of $MnCl_2$ and PMMA causes a decreased yield of monomer, the formation of relatively low-molecular-weight olygomers that contain anhydrides and esters, the evolution of several gases, and the formation of an insoluble product. These sealed tube reactions permit the detection and quantization of each of these species.

The TGA curves for PMMA are shown in Figure 3 (a). Kashiwagi et al. [24] showed that there is a direct correlation between the TGA curve and the method of preparation, presence of weak links, and end groups. From this curve, the material used in this study is anionacially polymerized with no irregular structures. Figure 3(b) shows the TGA curve for the $MnCl_2$-PMMA blend. The degradation can be separated into five distinct temperature regions; each will be discussed separately and the gases that evolve identified.

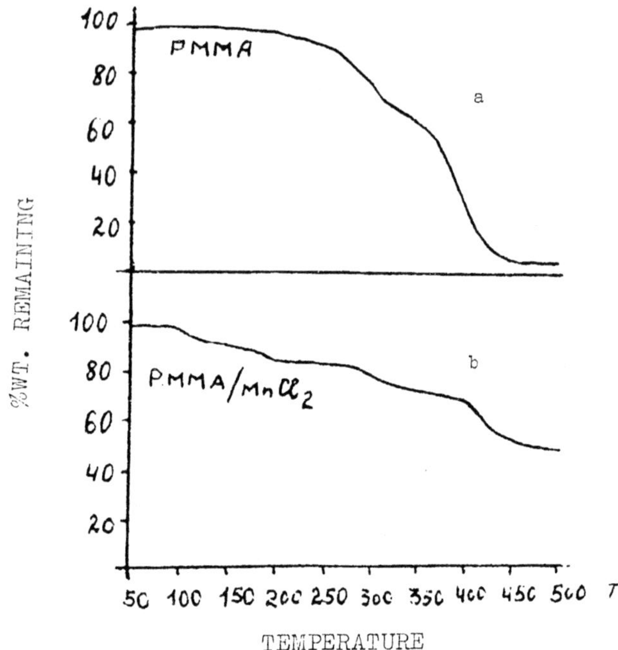


**Figure 3.** Thermogravimetric analysis curve for PMMA (top) and 1:1 blend of PMMA McCl2, run under an inert atmosphere at a scan rate of 20°C/min.

In the first region, 100-145°C, 7.5% of the sample volatilizes. The only material that is present in the FT-IR spectrum is water. Even though anhydrous $MnCl_2$ was used in these investigations, some water is picked up by the salt and is fairly easily lost.

An additional 9.5% of the sample volatilizes between 145 and 215°C. It is evident from the IR spectrum that this is due to monomer evolution. Note that in pure PMMA monomer evolution does not begin until higher temperature, and in sealed tube reactions no monomer is produced.

The third region of weight loss, 215-345°C, is characterized by the continued formation of monomer, as well as the appearance of methyl chloride and methanol (10.5% is lost in this region). Since the only source of chlorine is the $MnCl_2$, the presence of $CH_3Cl$ indicates that a reaction between $MnCl_2$ and PMMA has occurred. The formation of both methyl chloride and methanol must involve the methyl ester functionality of the PMMA.

Between 345 and 455°C, additional 19% of the sample volatilizes. In this region, the gases are identified as $CO_2$, CO, $CH_4$, HCl and an aliphatic acid.

The final region of weight loss, 455-630°C, accounts for 5% of the sample and is identified as HCl by infrared spectroscopy.

Scheme 4 presents an interpretation of the reaction and delineates a pathway whereby all of the products may arise.

$$\sim CH_2-\underset{\substack{|\\C=O\\|\\O\\|\\CH_3}}{\overset{CH_3}{C}}-CH_2-\underset{\substack{|\\C=O\\|\\O\\|\\CH_3}}{\overset{CH_3}{C}}\sim \xrightarrow{MnCl_2} \sim CH_2-\underset{\substack{|\\C=O\\|\\O\\|\\CH_3}}{\overset{CH_3}{C}}-CH_2-\underset{\substack{|\\C=O\\|\\O}}{\overset{CH_3}{C}}\sim \;+\;{}^{\cdot}CH_3\;+\;\sim CH_2-\underset{\substack{|\\C=O}}{\overset{CH_3}{C}}-CH_2-\underset{\substack{|\\C=O\\|\\O\\|\\CH_3}}{\overset{CH_3}{C}}\sim\;+\;{}^{\cdot}OCH_3$$

${}^{\cdot}CH_3 + {}^{\cdot}H \rightarrow CH_4 \qquad {}^{\cdot}CH_3 + {}^{\cdot}Cl \rightarrow CH_3Cl \qquad {}^{\cdot}OCH_3 + {}^{\cdot}H \rightarrow CH_3OH$

Degradation products (monomer) $\longleftarrow \sim CH_2-\overset{\overset{\textstyle CH_3}{|}}{\underset{\cdot}{C}}-CH_2-\overset{\overset{\textstyle CH_3}{|}}{\underset{\substack{|\\C=O\\\searrow\\\dot{O}\\|\\CH_3}}{C}}\sim \;+\;CO_2$

$\sim CH_2-\overset{\overset{\textstyle CH_3}{|}}{\underset{\cdot}{C}}-CH_2-\overset{\overset{\textstyle CH_3}{|}}{\underset{\substack{|\\C=O\\\searrow\\O\\|\\CH_3}}{C}}\sim \;+\;CO \longrightarrow$ Degradation products (monomer)

(Scheme 4 continues on next page)

McCl$_2$ facilitates the initial cleavage of carbon-oxygen, producing methoxy and methyl radicals, as well as carbonyl radicals and carboxyl radicals, along the polymer chain. These methoxy and methyl radicals may abstract protons and give methanol and methane, respectively. The methyl radical may also combine with a chlorine radical and produce methyl chloride. The carbonyl radical may decarbonylate, giving CO and a tertiary radical, while the carboxyl radical can decarboxylate, giving CO$_2$ and the same tertiary radical. These tertiary radicals will lead to PMMA degradation products, predominately monomer with some other smaller fractions. Since much more CO$_2$ is produced than CO, the loss of methyl radical must occur more often than the loss of methoxy radical. This carboxyl radical has an alternate pathway open. It may interact with the MnCl$_2$, producing an MnCl salt and a chlorine radical. This chlorine radical may combine with any other radical produced in the reaction to give products; methyl chloride, noted above, is one of those products. The manganese chloride salt of PMMA may lose a chlorine radical and produce a manganese radical on the PMMA chain. The combination of

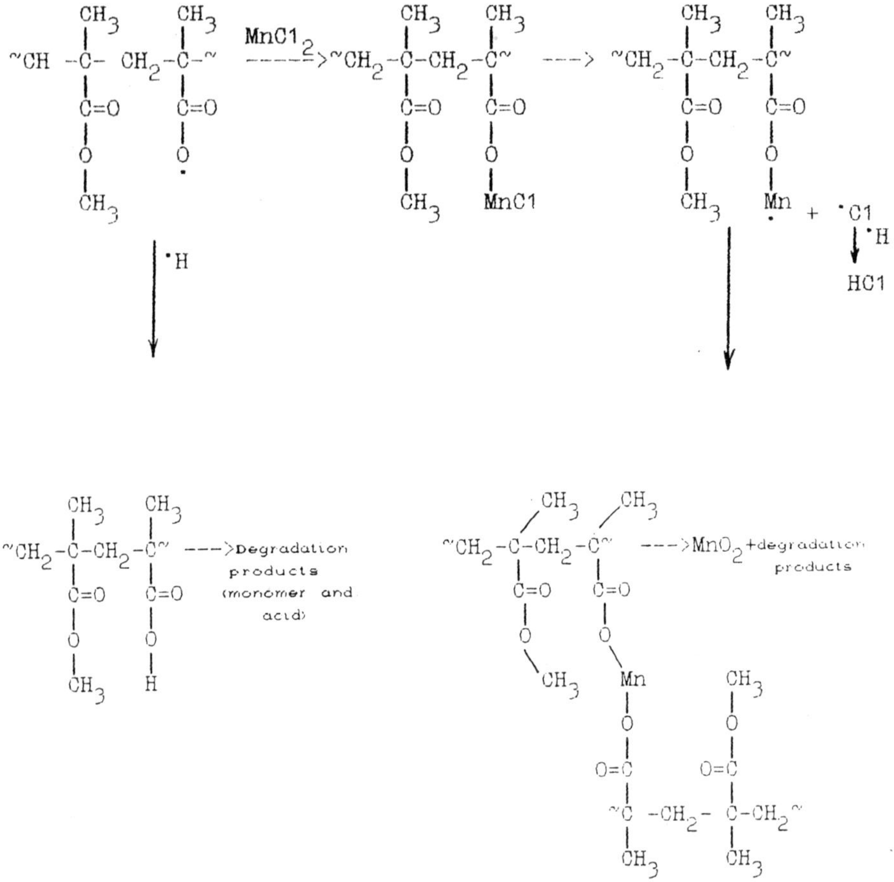

Scheme 4

this manganese radical with a second carboxyl radical gives the manganese ionomer of PMMA. This manganese ionomer may thermally decompose to produce $MnO_2$ and other degradation products. An aliphatic acid is also produced in the reaction. This may arise from the abstraction of a hydrogen by the carboxyl radical and subsequent depolymerization, giving monomer methacrylic acid as well as other degradation products.

Anhydride is also noted as a product of the reaction. Both intra- and intermolecular anhydride formation may occur, as shown in Scheme 5.

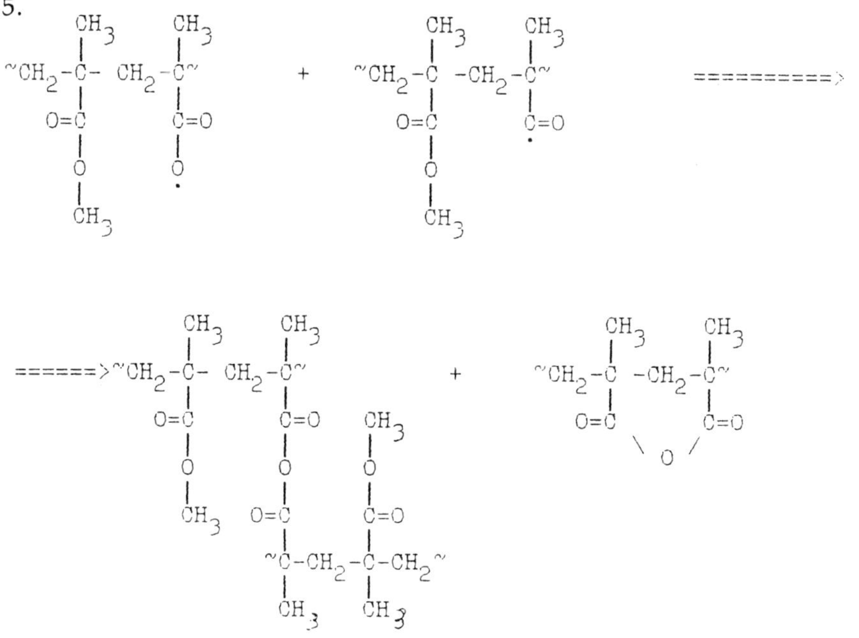

$$
\begin{array}{cc}
\quad CH_3 \quad\quad CH_3 & \quad CH_3 \quad\quad CH_3 \\
\sim CH_2-C-\ CH_2-C\sim \quad + \quad \sim CH_2-C\ -CH_2-C\sim & \quad\quad\quad =========\!\!> \\
\quad O=C \quad\quad C=O & \quad O=C \quad\quad C=O \\
\quad\quad O \quad\quad\quad O & \quad\quad O \\
\quad\quad CH_3 & \quad\quad CH_3
\end{array}
$$

$$
\begin{array}{cc}
======\!\!>\sim CH_2-C-\ CH_2-C\sim & \quad + \quad \sim CH_2-C\ -CH_2-C\sim \\
\quad O=C \quad\quad C=O \quad CH_3 & \quad O=C \quad\quad C=O \\
\quad\quad O \quad\quad\ O \quad\ O & \quad\quad\quad \backslash\ O\ / \\
\quad\quad CH_3 \quad O=C \quad\ C=O & \\
\quad\quad\quad \sim C-CH_2-C-CH_2\sim & \\
\quad\quad\quad CH_3 \quad\ CH_3 &
\end{array}
$$

Scheme 5

The limiting oxygen index [26] of the mixture of PMMA and manganese chloride has been investigated and indicates some efficacy for $MnCl_2$ as a flame retardant. At a 1:1 mass ratio the LOI, bottom ignition, [24], has increased from about 14 for pure PMMA to around 25 for the blend of $MnCl_2$ and PMMA. This loading is well beyond any value that might have commercial significance. It nonetheless shows that this additive has an effect.

This work must be contrasted to the investigation of the $ZnBr_2$-PMMA system by McNeill and McGuiness [15,16]. The gaseous products observed are identical for both additives; however, the order of appearance for these is not the same. With $ZnBr_2$, the initial product that is formed is methyl bromide, released at 160°C; the analogous methyl chloride is not formed with the manganese salt until temperatures above 220°C are reached. The initial reaction observed with $MnCl_2$ is the formation of monomer; indeed, this occurs at temperatures well below that where monomer is formed in pure PMMA.

On the other hand, with $ZnBr_2$, the initial reaction is elimination of methyl bromide with formation of the zinc salt of methacrylic acid. In both cases, the product of the reaction is a metal salt of methacrylic acid. In the work [22], this has been categorized as an ionomer to indicate the stability expected for this species. Considering the great similarity between these two reactions, it seems very likely that the reactions are quite similar; the only likely difference is in the interpretation of the results.

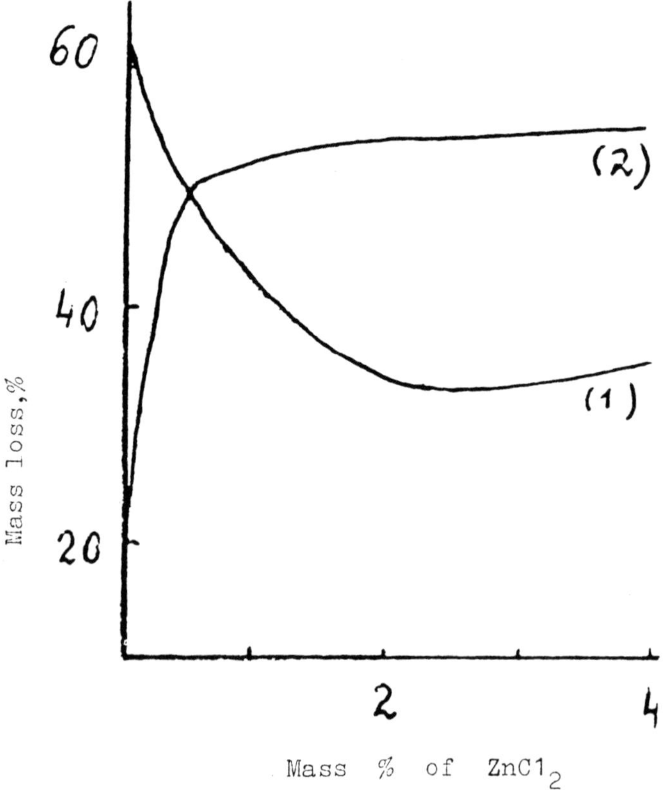


**Figure 4.** Mass loss of PMMA and PMA vs $ZnCl_2$ concentration (300°, in vacuo, 2 hr): 1-PMMA, 2-PMA.

So, there are both advantages and disadvantages to the use of manganese chloride as a flame retardant for PMMA. On the one hand, $MnCl_2$ promotes an initial depolymerization of the polymer and thus promotes burning. On the other hand, a nonvolatile ionomeric species is ultimately produced. Unfortunately, there is sufficient degradation of the polymer to generate burnable materials so this species is not an adequate flame retardant.

Effect of zinc chloride on thermal decomposition of polymethacrylic and polyacrylic esters studied in the work [27].

Polymethacrylates and polyacrylates decompose by different mechanisms, the former depolymerizing almost completely, and the latter decompositing by decomposition at the ester groups.

**Table 2.** The main products of PMMA thermal decomposition (mass % of the amount of the starting polymer (at 300° (2 hr) in the presence of $ZnCl_2$.

| $ZnCl_2$, mass% | 0 | 0.05 | 0.5 | 2 | 10 |
|---|---|---|---|---|---|
| Extent of decomposition | 60.7 | 55.6 | 44.8 | 30.0 | 25.0 |
| Solid residue | 39.3 | 44.4 | 55.2 | 70.0 | 75.0 |
| Chain fragments | 0.2 | 1.3 | – | 1.9 | 1.9 |
| MMA | 60.2 | 54.0 | 43.0 | 25.3 | 11.5 |
| $CH_3OH$ | 0.40 | 0.34 | 0.40 | 0.90 | 2.60 |
| $H_2O$ | 0 | 0 | 0 | 0 | 3.10 |
| $CH_3Cl$ | 0 | 0.03 | 0.28 | 1.29 | 2.64 |
| $CO_2$ | 0 | 0.12 | 0.15 | 0.25 | 0.20 |
| Other products (CO, $CH_4$, $C_3H_6$) | 0 | 0.020 | 0.03 | 0.1 | 1.14 |
| Balance | 100.1 | 100.1 | 99.06 | 99.75 | 98.08 |

**Table 3.** The main products of PBMA thermal decomposition (mass% of the amount of the starting polymer) at 300° (2 hr) in the presence of $ZnCl_2$

| $ZnCl_2$, mass% | 0 | 0.5 | 3 |
|---|---|---|---|
| Extent of decomposition | 56.5 | 58.0 | 33.0 |
| Solid residue | 43.5 | 42.0 | 67.0 |
| Chain fragments | 2.0 | 4.3 | – |
| BMA | 49.0 | 43.5 | 23.0 |
| n-$C_4H_9OH$ | 4.0 | 7.5 | 6.4 |
| $H_2O$ | 0.25 | 0.39 | 0.60 |
| $CO_2$ | 0.17 | 0.24 | 0.20 |
| $C_4H_9Cl$ | 0 | 0.14 | 0.35 |
| Butylene | 1.0 | 0.9 | 0.6 |
| Other products ($H_2$, $CH_4$, $C_3H_6$, CO) | 0.1 | 0.9 | 0.6 |
| Balance | 100.02 | 99.87 | 98.75 |

**Table 4.** The main products of PMA thermal decomposition (mass% of the amount of the starting polymer) at 300° (2 hr) in the presence of $ZnCl_2$

| $ZnCl_2$, mass% | 0 | 0.05 | 0.5 | 2 | 10 |
|---|---|---|---|---|---|
| Extent decomposition | 15.6 | 32.0 | 47.0 | 43.5 | 49.0 |
| Solid residue | 84.4 | 68.0 | 53.0 | 56.5 | 51.0 |
| Chain fragments | 12.3 | 16.8 | 20.0 | 17.8 | 16.0 |
| $CH_3OH$ | 2.6 | 13.6 | 20.4 | 16.6 | 17.6 |
| MMA | traces | 0.2 | 0.3 | 0.3 | 0.4 |
| $H_2O$ | 0.1 | 0.2 | 0.3 | 0.35 | 1.0 |
| $CO_2$ | 0.3 | 1.1 | 5.0 | 7.0 | 7.7 |
| $CH_3Cl$ | 0 | 0.04 | 0.4 | 1.1 | 5.1 |
| Other products ($H_2$, $CH_4$, CO) | 0 | traces | 0.3 | 0.4 | 1.1 |
| Balance | 99.7 | 99.94 | 99.6 | 100.05 | 99.9 |

**Table 5.** The main products of PBA thermal decomposition (mass % of the amount of the starting polymer) at 300° (2 hr) in the presence of $ZnCl_2$

| $ZnCl_2$, mass% | 0 | 0.5 | 0.5 | 2 |
|---|---|---|---|---|
| Extent of decomposition | 21.8 | 14.3 | 20.6 | 20.2 |
| Solid residue | 78.2 | 85.7 | 79.4 | 79.8 |
| Chain fragments | 9.5 | 2.0 | 2.0 | 1.0 |
| n-$C_4H_9OH$ | 9.0 | 10.0 | 16.0 | 15.5 |
| $H_2O$ | 0.4 | 0.2 | 0.3 | 0.5 |
| BMA | traces | | | |
| $CO_2$ | 0.8 | 0.8 | 0.8 | 1.2 |
| Butylene | 1.4 | 1.3 | 1.0 | 1.5 |
| $C_4H_9Cl$ | 0 | traces | 0.1 | 0.3 |
| Other products ($H_2$, $CH_4$, CO) | 0.1 | 0.1 | 0.1 | 0.2 |
| Balance | 99.4 | 100.1 | 99.7 | 100.0 |

For PMMA the yield of monomer falls, indicating suppression of the depolymerization; the yields of alcohol and chain fragments increase and $CO_2$ rises (Table 2). Similar effects are observed for poly-n-butylmethacrylate (PBMA) (Table 3). For thermal decomposition of PMA (Table 4) the yields of alcohol and $CO_2$ increase sharply even for small additions of $ZnCl_2$. Similar effects are observed for poly-n-butylacrylate (PBA) (Table 5).

Alkyl chloride was quantitatively determined in the products of thermal decomposition for both polymethacrylic and polyacrylic esters in the presence of $ZnCl_2$.

Experimental data suggest that $CH_3Cl$ is formed mainly by intramolecular reaction of $ZnCl_2$ with ester carbonyl groups.

Table 6 shows that the formation of alkyl chlorides from poly-acrylic polymers in the presence of $ZnCl_2$ is a general reaction, its rate being reduced with increases of the length of hydrocarbon radical in the ester group. $ZnCl_2$ affects similarly the formation of insoluble fractions during PMMA and PMA thermal decomposition. Assuming the same mechanism and rates of $ZnCl_2$ interaction with the ester groups upon formation of alkyl chlorides, it is possible that crosslinking proceeds mainly by molecular interaction of ester groups with $ZnCl_2$. The contribution of this reaction, as compared to the intramolecular reaction mentioned above, is small; therefore it results in complete loss of solubility of the polymeric residue only for large amounts of $ZnCl_2$.

**Table 6.** Formation of alkyl chlorides (mass%) as a function of $ZnCl_2$ content during thermal decomposition of polymethacrylates and polyacrylates (300°, in vacuo, 2 hr)

| $ZnCl_2$, mass% | $CH_3Cl$ | | $C_4H_9Cl$ | |
|---|---|---|---|---|
| | PMMA | PMA | PBMA | PBA |
| 0.05 | 0.03 | 0.04 | 0.03 | 0.03 |
| 0.5 | 0.3 | 0.4 | 0.1 | 0.1 |
| 2 | 1.3 | 1.2 | 0.4 | 0.3 |
| 10 | 2.6 | 5.1 | – | – |

As confirmed by thermovolumentric analysis [28] (Figure 5) the depolymerization is suppressed by $ZnCl_2$. In the presence of $ZnCl_2$, the peak at 250°, corresponding to the initiation step of the thermal decomposition with the unsaturated end-groups of the macromolecules, is reduced considerably. The peak associated with the random

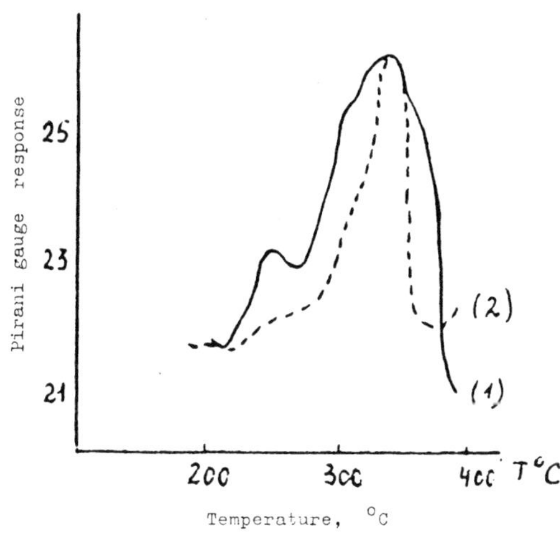


**Figure 5.** TVA curves for pure PMMA and in the presence of $ZnCl_2$ (2 mass%): 1-PMMA, 2-PMMA with $ZnCl_2$.

initiation step of decomposition remains unchanged. The formation of $ZnCl_2$ complex with a chain-end macroradical yields alcohol and inhibits depolymerization.

This process may be described by scheme 6.

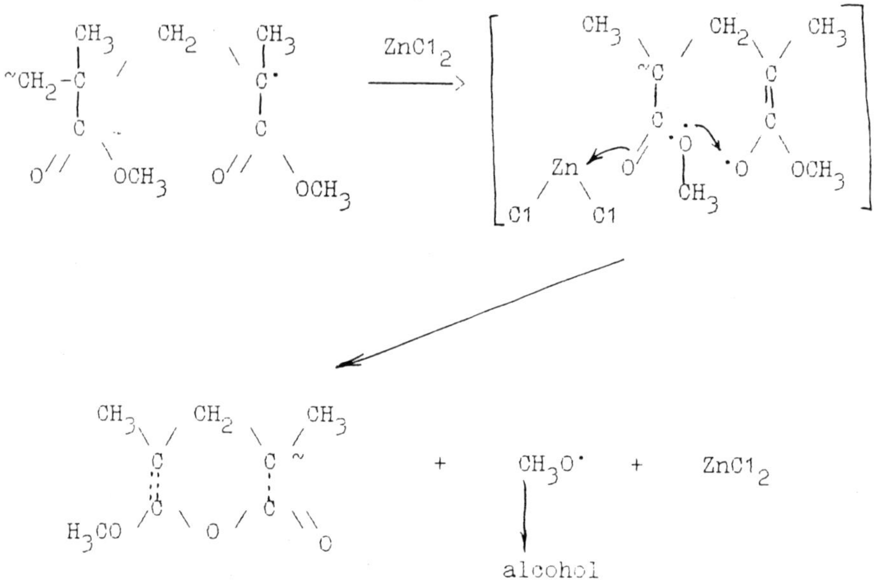

Scheme 6.

IR-spectra of PMMA samples decomposed in the presence of $ZnCl_2$ show absorption at 1760 $cm^{-1}$ corresponding to $\gamma$-$\delta$-unsaturated lactones.

# TRANSITION-METAL ACETYLACETONATES

The main effects of the additive result from complex formation with the unsaturated chain ends and from attack of acetylacetonate radicals on the polymer, leading to some low temperature evolution of monomer, but also to stabilization of the PMMA in the normal temperature region for decomposition as a result of the formation of some anhydride rings.

Polymer chemists have become interested in transition metal acetylacetonates (acacs) for various reasons. They have been used as initiators in polymerization, they have been reported to show pro and antioxidant activity in polyolefins and they have also been suggested as fire retardants and smoke suppressants for certain polymers [29-33].

McNeill and Liggat in the work [34] studied the effect of small amounts of cobalt (111) acetylacetonate on thermal degradation behavior of PMMA. Authors suggested that Co (acac3) would significantly modify the thermal degradation behavior of the polymer, by complexing with the ester groups of PMMA. The composition range examined was from one part of the additive to 400 monomer units up to one part to 10 monomer units. The structure of Co (acac3) is represented by formula

$$Co^{3+}\left[\ \begin{array}{c} O=C\diagdown^{CH_3} \\ \diagup \quad CH \\ O-C\diagup \\ \diagdown_{CH_3} \end{array}\ \right]_3$$

Thermal degradation was studied by thermal volatilization analysis (TVA) [35,36]. This approach permits to examine separately the four main product fractions: non-condensable gases, condensable gases and volatile liquids, cold ring fraction (tars, waxes, etc.), residue at any stage of degradation.

Figure 6 shows TVA curves for three PMMA samples used in this investigation. There is two peaks corresponding to two stages of degradation initiated at unsaturated ends and by random scission, respectively. The sizes of two peaks depend on the MW of the polymer. For low MW (zip length) the peaks are of similar size. As the MW becomes greater than the zip length, the first peak becomes smaller than the second. For very high MW ($\gg$ zip length) the first peak is very small. These differences in the size of the first TVA peak are clearly illustrated in the TVA curves for the three PMMA samples, shown in Figure 6.

PMMA behaves differently if a complexing agent is present, such as a transition metal halide or acetate. For example, with $ZnBr_2$ present, methyl bromide and methanol become important degradation products and less monomer is evolved [37,38]. These changes result from the formation of the complex with structure

$$\sim CH_2\diagdown \overset{CH_3}{\underset{|}{C}}\diagup CH_2\diagdown \overset{CH_3}{\underset{|}{C}}\sim$$
$$H_3C-O \diagup \overset{\diagdown\diagdown}{O}\diagdown \underset{Zn}{\diagup}\diagup O \diagdown O-CH_3$$
$$Br \diagup \qquad \diagdown Br$$

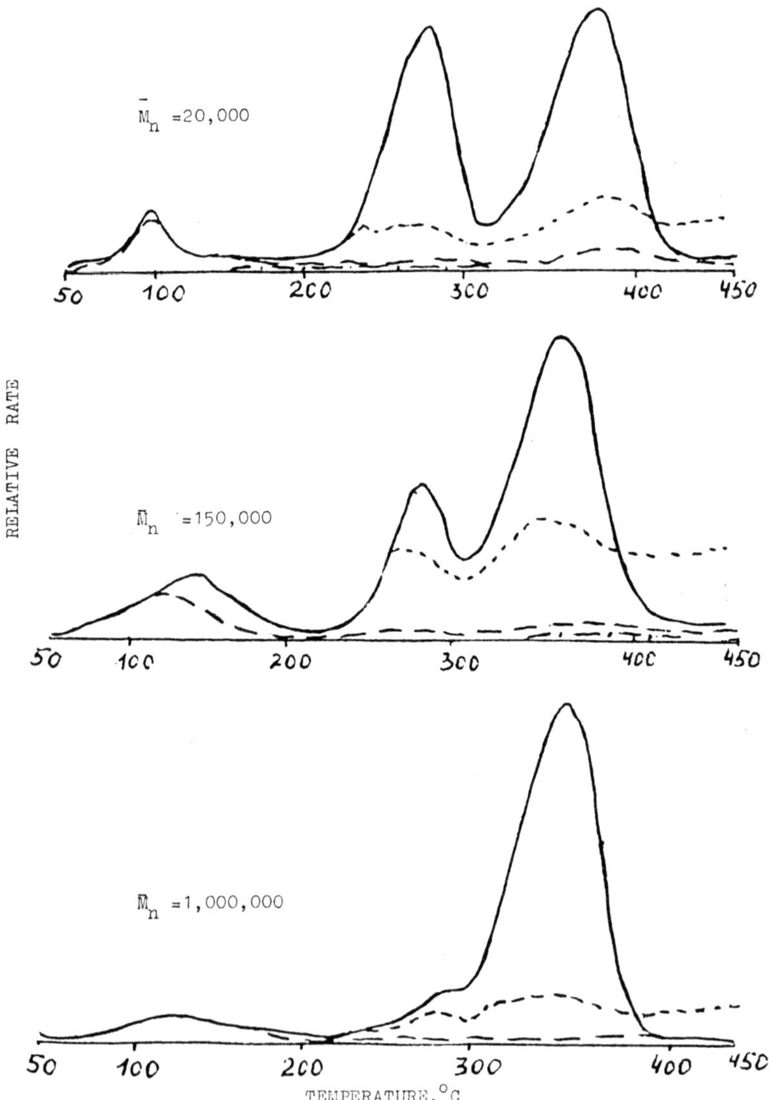


**Figure 6.** TVA curves (vacuum, 10°/min) for PMMA samples of different molecular weights. Key to traces: ______0°, - - - - -75°C, _ _ -100°C, -196°C.

Suitable metal acacs might be expected to complex with the ester side groups in PMMA. This could lead to differences in the stability and degradation mechanism for the polymer.

The main reaction is:

$$Co(acac)_3 \rightarrow Co\,(acac)_2 + acac^{\bullet}$$
$$\rightarrow \text{acetylacetone}$$

UV spectra of the blend Co (acac)$_3$-PMMA show that there is no complex formed with the ester groups. On the other hand, if Co (acac)$_2$ is blended with PMMA, there is a shift of carbonyl absorption for the acac salt from $\lambda_{max}$ at 291 nm to 287 nm. Thus during degradation of Co (acac)$_3$-PMMA blend, if the Co(acac)$_3$ decomposes to Co (acac)$_2$, complex formation may then occur involving the ester side groups.

The schematic TVA curve shown in Figure 7.

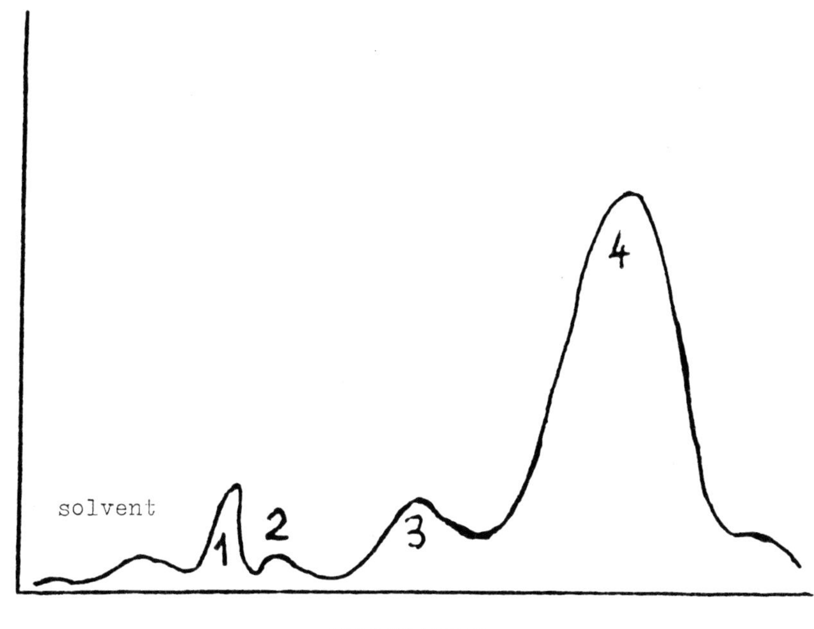


**Figure 7.** Schematic diagram of the four stages of degradation evident in the TVA curves for Co (acac)$_3$-PMMA blends.

Stage 1: $T_{max}$ about 135°C. The peak height is inversely proportional to MW and independent of [Co(acac)$_3$], in the range down even to 1:400. Due mainly to monomer. The stage 3 unsaturated end-initiated reaction decreases as the stage 1 peak increases. Authors concluded that at this stage, end-initiated depolymerization of limited zip length occurs, due to Co (acac)3 forming a $\pi$-complex at unsaturated chain ends (average of 1 end of this type per 400 monomer units in low MW sample).

Stage 2: $T_{max}$ about 190°C. Peak height varies directly with both the MW of the polymer and [Co(acac)3]. Prominent only at high MW and high [Co(acac)3]. Due mainly to monomer. Traces of non-condensable gas also formed when [Co(acac)3] is high. At this stage, depolymerization with longer zip length occurs, plus some other process.

Stage 3: $T_{max}$ about 270°C. Peak height reduced in all blends compared with that for pure PMMA of the same MW. Due mainly to monomer. Despite stage 1, end-initiated depolymerization still occurs. Unsaturated chain ends are still present or can be regenerated during stage 1. Zip length is reduced.

Stage 4: $T_{max}$ 360-380°C. Peak shifts to slightly higher temperature in every case and is accompanied by some production of gases which are noncondensable. Authors interpreted this stage by reduce of zip length and some side reaction occurring along with depolymerization.

The various products, in addition to monomer, investigated by IR and UV spectra. These results are listed in Table 7.

**Table 7.**   Volatile products evolved during the degradation of Co(acac)3-PMMA blends, in addition to monomer

| Up to end of Stage 2 | Up to end of Stage 4 |
|---|---|
| Acetylacetone | Acetylacetone |
| Methanol (above 200°C) | Methanol<br>Carbon dioxide<br>A ketone<br>Methyl acetate<br>Cyclic anhydride<br>Carbon monoxide<br>Isobutene (trace)<br>Dimethyl ether (trace) |

Authors investigated also IR and UV spectra of film (1:50 blend), heating to 500°C.

The various features of the degradation of PMMA in the presence of Co(acac)3 as the temperature is gradually increased can be considered as following.

A. At the lowest temperatures (around 100°C), degradation to monomer commences as a result of decomposition of the $\pi$-complex formed between the acac salt and unsaturated chain ends in the polymer (Scheme 7).

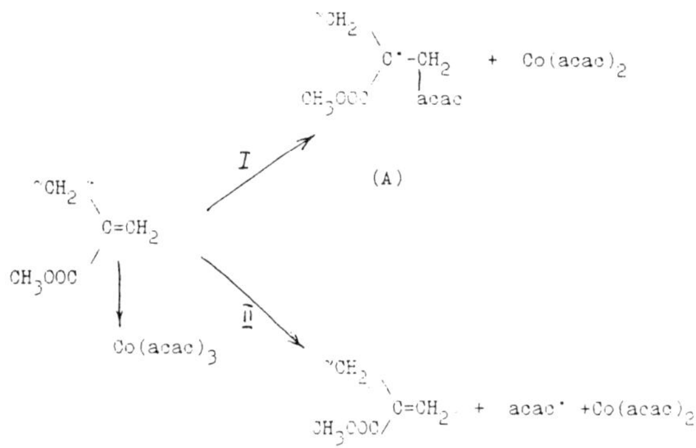

Scheme 7.

Scheme 8.

In route 1, depolymerization of the radical (A) need not remove all the unsaturated chain ends if the zip length is less than the chain length: short zip length is likely in the temperature region not far above the glass transition temperature.

In route 11 unsaturated ends remain and depolymerization is initiated by acac radical attack on the chain.

B. Thermal decomposition of Co (acac)$_3$ produces Co(acac)$_2$ and acac radicals. The radicals initiate depolymerization by H-abstraction: the unchain macroradical initially formed undergoes scission to give a terminal radical which can unzip. The Co(acac)$_2$ complexes with ester side groups; this promotes some side group scission leading to methanol and CO as products and generating double bonds in the backbone at these points (Scheme 8). The complex can also decompose to form cobalt carboxylate structures (Scheme 9). These structures block unzipping.

C. $CH_3O$ radicals (from side group scission) also attack some ester groups forming anhydride rings (Scheme 10).

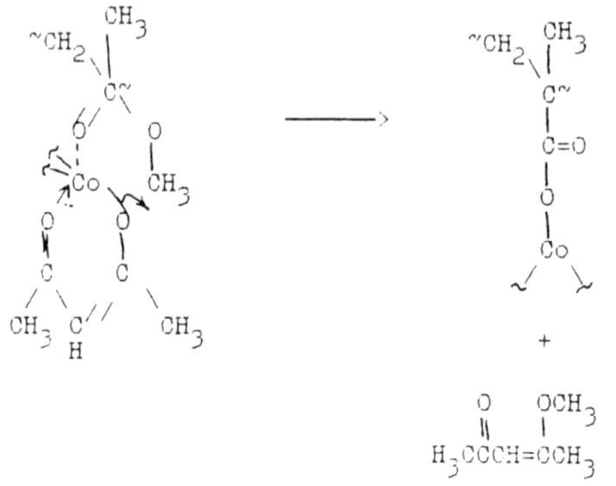

Scheme 9.

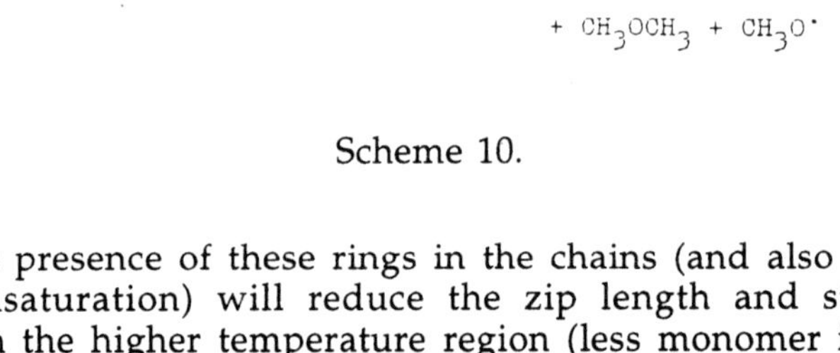

Scheme 10.

D. The presence of these rings in the chains (and also carboxylate salt or unsaturation) will reduce the zip length and stabilize the polymer in the higher temperature region (less monomer is produced from initial scission). Hence the upward shift in the last $T_{max}$, and also in the stage 3 $T_{max}$ if evident.

E. Fragmentation of the modified polymer structure (with some rings and unsaturation) gives mainly monomer, but also small amounts of products related to these abnormal structures (Scheme 11).

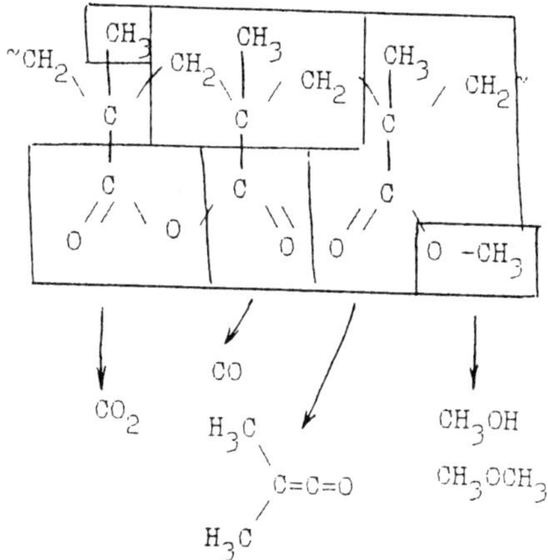

Scheme 11.

McNeill and Liggat [39] also examined of the effect of small amounts of manganese (111) acetylacetonate on the thermal stability and degradation mechanism of radically polymerized poly (methylmethacrylate), using a similar method of the work [22].

Thermal degradation was studied by thermal volatilization analysis under vacuum and by thermogravimetry and differential scanning calorimetry (DSC) under a flow of nitrogen. In each case, programmed heating at 10°C per minute was employed.

The TVA approach permits the four main product fractions (non-condensable gases, condensable gases and volatile liquids, cold ring fraction of tar and was products and residue) to be examined separately.

The DSC curve for Mn(acac)3 (Figure 8) shows a very characteristic combination of irreversible endotherm-exotherm pairs at 170, 175°C and 257,260°C. The initial pair has been ascribed to the homolytic scission of a single acac ligand from Mn(acac)3, with the subsequent regeneration of acetylacetone and the formation of the bis-chelate [40].

The first stage of the TG curve (Figure 9) is in agreement with this interpretation, since the initial weight loss (24%) corresponds to loss of one ligand. The subsequent pattern of weight loss is more complex than expected, however, suggesting that other fragmentation reactions may also be occurring. Other small radical species such as acetyl may be produced during this fragmentation. The ultimate residue is the metal oxide.

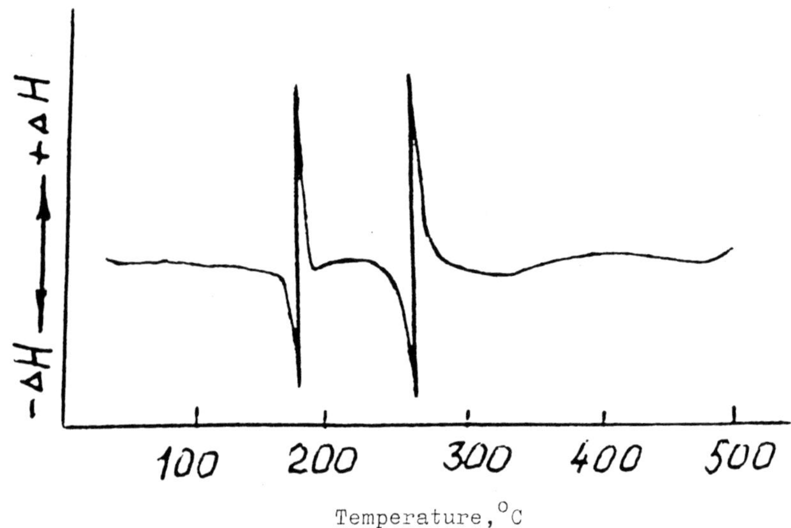



**Figure 8.** DSC curve (dynamic nitrogen atmosphere, 10°C/min for Mn(acac)3.

The decomposition of Mn(acac)3 may therefore be summarized as follows:

$$Mn(acac)_3 \longrightarrow Mn(acac)_2 + acac\cdot$$

Complex decomposition

Acetylacetone via H-abstraction

This is similar to the decomposition of the Co (111) chelate [34] and suggests that the chemistry of interaction in the thermal degradation of PMMA containing the Mn (111) salt may be basically similar.

The temperatures of onset of degradation and maximum rate for the main decomposition of Mn(acac)3-PMMA blends compared with that of the pure polymer shows in Table 8.

**Table 8.** Thermal stability of Mn(acac)3-PMMA blends compared with that of the pure polymer

|  | High MW | | | Medium MW | | |
|---|---|---|---|---|---|---|
|  | Alone | 1:50[a] | 1:10[a] | Alone | 1:50[a] | 1:10[a] |
| Onset of degradation (°C) | 210 | 130[b] | 125[b] | 220 | 130[b] | 130[b] |
| $T_{max}$ of main decomposition (°C) | 350 | 365 | 400 | 360 | 365 | 400 |

[a] Ratios are acac salt/monomer unit.
[b] Ignoring solvent evolution.

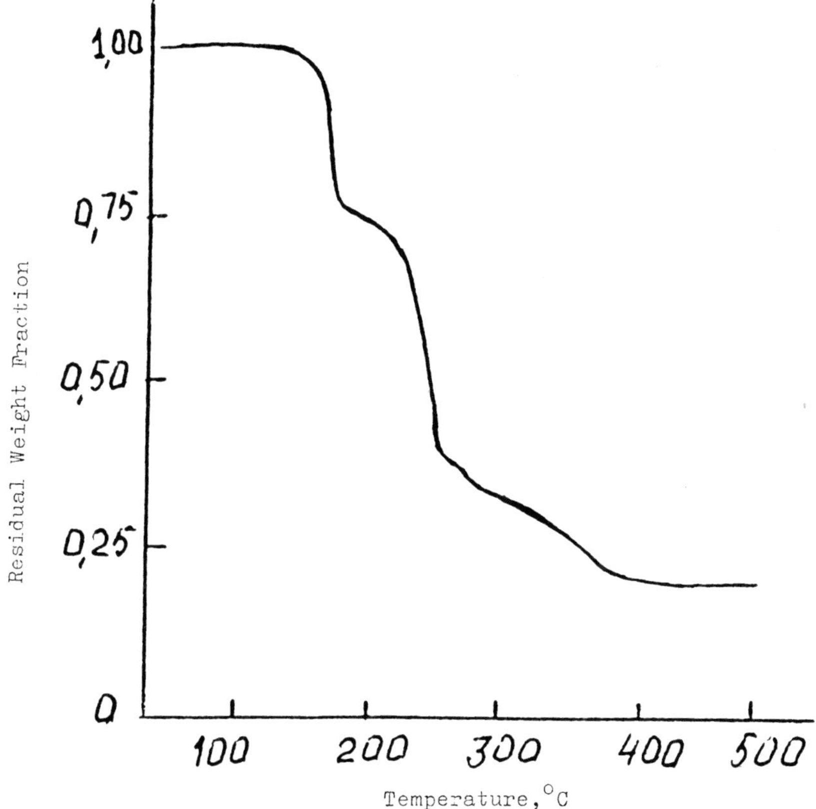


**Figure 9.** TG curve for Mn(acac)3 under dynamic atmosphere at 10°C/min.

Pure PMMA, when heated at 10°C/min, shows no volatilization below 210°C, except for the release of residual solvent. In contrast, the blends with Mn(acac)3 display evidence of degradation at temperatures as low as 125°C.

The features observed in Table 8 can be divided into two categories:

1. new low temperature degradation processes;
2. stabilization of the main degradation reaction.

In these respects, the Mn(acac)3 blends show similar TVA behavior to the Co(acac)3 blends previously studied.

The volatile products identified at various stages of the reaction are listed in Table 9.

**Table 9.**   Volatile products evolved during the degradation of Mn(acac)3-PMMA blends, in addition to monomer[a].

| Up to end of stage 1 | Up to end of stage 2 | Up to end of stage 3 |
|---|---|---|
| Acetylacetone | Acetylacetone<br>Methanol<br>Acetone<br>Methyl acetate<br>Carbon dioxide | Acetylacetone<br>Methanol<br>Carbon dioxide<br>Dimethyl ether<br>Butene<br>Isobutene<br>Dimethyl ketene<br>Acetone<br>Methyl acetate<br>Carbon monoxide<br>Methane |

[a] See Figure 10 for key to stages of degradation

Monomer was by far the most important product at every stage. With the exception of acetone, an observed decomposition product of Mn(acac)3, the additional products are the same as those identified from the Co(acac)3-PMMA blends. Their amounts, however, tended to be higher, and acetylacetone first appeared at lower temperatures.

The three main phases of decomposition of Mn(acac)3-PMMA blends authors describes by the schematic diagram (Figure 10).

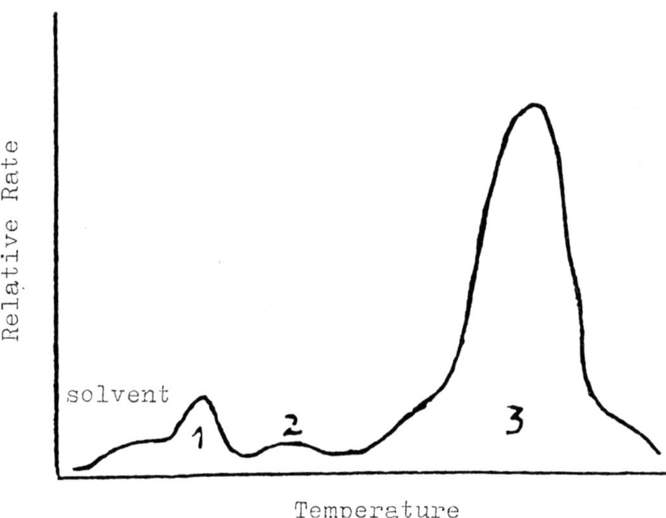


**Figure 10.**   Schematic diagram indicating the three stages of degradation for which the degradation products are listed in Table 9.

At the onset of degradation, phase 1, two closely overlapping processes around 15-160°C are associated with the production of acetylacetone and a small amount of monomer, as a result of decomposition of $Mn(acac)_3$ and initiation of depolymerization of the polymer by attack of acac• radicals (H-abstraction) and consequent chain scission to give a macroradical which can depropagate to monomer.

The fact of present of acetone and carbon dioxide as products at stage 2 identify this phase of the degradation as related to the decomposition of the bis-chelate formed in stage 1.

Decomposition of the complex between $Mn(acac)_2$ and ester side groups can give rise to carboxylate structures which at stage 3 can result in blocking of unzipping. It is for this reason that the normally pronounced end initiated unzipping process of medium MW PMMA is so effectively inhibited in the case of the $Mn(acac)_3$ blend. Authors concluded firstly, that the degradation behavior of $Mn(acac)_3$ blends with PMMA is very similar to that of the $Co(acac)_3$ blends despite the difference that only the Mn(111) chelate can complex with the ester groups. It is concluded secondly, that the effect of this type of complex formation on acac ligand scission at the lowest degradation temperatures is slight and that the effect of the bis-chelate, formed by decomposition of the tris-chelate in each case, much more important.

# REFERENCES


1.  **C.F. Cullis and V.V. Hirschler,** *The Combustion of Organic Polymers,* Oxford University Press. Oxford 1981.
2.  **M.M. Hirschler,** In *Development in Polymer Stabilization - 5/*edited by G. Scott. p. 107, Applied Science Publishers, London 1981.
3.  **R.M. Aseeva, G.E. Zaikov,** *Makromol. Chem., Macromol. Smp.,* 1993, V. 74, p. 335.
4.  **R.M. Aseeva, G.E. Zaikov,** *Combustion of Polymer Materials,* Hanser Publishers, N.-Y., 1981.
5.  **G.-T. Cardenas and C.-C. Retamal,** *Termochimica Acta,* 1991, V. 176, p. 233.
6.  **G.-T. Cardenas, K. Klabunde and E. Dale,** *Langmuir,* 1987, v. 3, p. 986.
7.  **G.-T. Cardenas, K. Klabunde and E. Dale,** *Proc. Opt. Eng., SPEI,* 1987, V. 821, p. 206.
8.  **G.-T. Cardenas, C.-C. Retamal,** *XVIII Jornadas Chilenas de Quimica,* Nov. 1989 Santiago (Chile), p. 350.
9.  **S.J. Sirdesai and C.A. Wilkie,** *J. Appl. Polym. Sci.,* 1989, v. 27, p. 863.

10. S.J. SIrdesai and C.A. Wilkie, *J. Appl. Polym. Sci.*, 1989, v. 37, p. 1595.
11. C.A. Wilkie, S.J. Sirdesai, T. Suebsaeng and P. Chang, *Fire Safety J.*, 1989, v. 15, p. 297.
12. R.S. Beer, C.A. Wilkie and M.L. Mittleman, *J. Appl. Polym. Sci.*, 1992, v. 46, p. 297.
13. D. Kiraly, K. Zalatnai and M.T. Beck, *J. Inorg. Nucl. Chem.*, 1959, v. 11, p. 170.
14. L. Gmelin, In *Gmelins Handbuch der Anorganische Chemie*, Verlag Chemie, Berlin, vol. cr., pt. B, 1962, p. 208.
15. I.C. McNeill and R.C. McGuiness, *Polym. Degrad. and Stabil.*, 1984, v. 9, p. 167.
17. C.A. Wilkie, J.W. Pettegrew and C.E. Brown, *J. Polym. Sci. Polym. Lett. Ed.*, 1981, v. 19, p. 409.
18. L.E. Manring, *Macromolecules*, 1988, v. 21, p. 528.
19. L.E. Manring, *Macromolecules* , 1989, v. 22, p. 2673.
20. L.E. Manring, D.Y. Sogah and G.M. Cohen, *Macromolecules*, 1989, v. 22, p. 4652.
21. L.E. Manring, *Macromolecules*, 1991, v. 24, p. 3304.
22. C.A. Wilkie, J.T. Leone, M.L. Mittleman, *J. Appl. Polym. Sci.*, 1991, v. 42, p. 1133.
23. R.H. Pierson, A.N. Fletcher, and E.S. Gantz, *Anal. Chem.*, 1956, v. 28, p. 1218.
24. T. Kashiwagi, A. Inaba, J.E. Brown, K. Hatada, T. Kitayama and E. Masuda, *Macromolecules*, 1986, v. 19, p. 2160.
25. H.R. Allcock and F.W. Lampe, *Contemporary Polymer Science*, Prentice-Hall, Englewood Cliis, New Jersey, 1981, p. 378-394.
26. D.E. Stuetz, A.H. Diedwardo, F. Zitomer, and B.P. Barnes, *J. Polym. Sci., Polym. Chem. Ed*, 1975, v. 13, p. 585.
27. L.S. Kochneva, N.A. Koplova, L.M. Terman and Yu.D. Semchikov, *European Polym. J.*, 1979, v. 15, p. 575.
28. S.L. Madorsky, *Thermal Degradation of Organic Polymers*, John Wiley and Sons, N.-Y., 1964.
29. E.M. Arnett, M.A. Mendelsohn, *J.Am. Chem. Soc.*, 1962, v. 84, p. 3821, 3824.
30. T. Otsu, N. Minimii, Y. Nishikawa, *J. Macromol. Sci.-Chem.* 1968, A2, p. 983.
31. H.S. Laver, *In Developments in Polymer Stabilization*, V.1, ed. Scott G., Applied Science, London, 1977, p. 167.
32. M.U. Amin, G. Scott, *Eur. Polym. J.*, 1974, v. 10, p. 1019.
33. N.J. Kroenke, *J. Appl. Polym. Sci.*, 1981, v. 26, p. 1167.
34. I.C. McNeill, J.J. Liggatt, *Polym. Degrad. and Stabil.*, 1990, v. 29, p. 93.
35. I.C. McNeill, *Eur. Polym. J.*, 1967, v. 3, p. 409.
36. I.C. McNeill, *Eur. Polm. J.*, 1970, v. 6, p. 373.

37. **I.C. McNeill**, *R.C. McGuiness*, Polym. Degrad. and Stabil., 1984, v. 9, p. 167.
38. **I.C. McNeill**, *R.C. McGuiness*, Polym. Degrad. and Stabil., 1984, v. 9, p. 209.
39. **I.C. McNeill**, *J.J. Liggat, Polym.* Degradation Stab., 1992, v. 37, p. 25.
40. **I. Yoshida, H. Kobayashi, K. Ueon,** *Bull. Chem. Soc. Japan,* 1974, v. 47, p. 2203.

# Fire and Heat Shield Materials, Based on Sulfochlorinated Polyethylene


*M.A. Shashkina*, R.M. Aseeva,*
*A.A. Donskoy* and G.E. Zaikov*
**Institute of aviation materials, Moscow*
*Institute of Chemical Physics, Russian Academy of Sciences*
*117334, Moscow, Kosygin Str., 4, Russia*


## 1. Introduction

The increase of aircraft safety is connected with objective, detailed study of accident causes. For this purpose the analysis of flight parameters and crew working during start, flight and accident is performed. Objective witnesses are flight information recording systems (FIRS), recording the main parameters of a flight on tape. That is why FIRS are placed usually into metal container with fire shield cover, intended to block and decrease heat flow from the surroundings. Container with fire shield cover should protect information recording from thermal degradation in case of fire zone entering as well as from mechanical destruction after falling from a height, from humidity influence at water entering. This so-called "black box" gives the possibility to clear up objective picture of an accident happened. Recently such information recording systems are placed at railway transport, water and underground transport.

The present paper considers physico-chemical aspects of creation of efficient fire and heat shielding materials (FHSM) for FIRS. The data on working out FHSM with high exploitational characteristics, based on sulfochlorinated polyethylene, are shown. Great attention is paid to the analysis of fire shield action of the material worked out.

## 2. PHYSICO-CHEMICAL ASPECTS OF CREATION OF EFFICIENT FHSM FOR FLIGHT INFORMATION RECORDING SYSTEMS

There are several types of FIRS, differing by construction and amount of the information recorded. The limit of acceptable temperature inside container (for example, 60°C for Soviet equivalent of Darcon and 350°C - for metal tape) changes in dependence on material nature, used for tape preparing.

First constructions of FIRS foresaw external heat shielding cover, metal blow resistant cover, inside which heat isolation layer was placed. As it is known, fiber glasses possess high heat shielding properties. However, their blow resistance is not high. If aircraft falls down, such covers break. The necessity there appears to create elastic blow resistant covers, providing safe heat shield of the device and after its falling from height. Moreover, the task of decrease of device overall size and its weight put forward the question of working out highly efficient FHSM, allowing protection of the information carrier without using internal heat shielding cover.

Great experience in creation of heat shield materials, including elastic ones, was accumulated in missile-space technics. However, FHSM for FIRS differ from traditional heat shielding for air-space vehicles by working conditions, especially temperature (it is comparitively low 1100-1600°C) as well as by time of heat influence (long duration - up to 30 minutes).

FHSM work is composed from two stages: active and passive. Active stage is deterministic for traditional heat shield. In case of FIRS active stage is stipulated by flame influence on cover surface during a long time. After the end of external heat influence passive stage starts: the cover, stored heat energy during active stage, becomes itself the source of heat.

For efficient work of FHSM it is important to lose maximum amount of feeding energy for material transformation during the first stage, and accumulate minimum of it. At the second stage of work the cover material should provide aimed heat emission into the surroundings.

The literature on fire shield describes, on the whole, designs including FHSM, internal heat shielding and radiation energy reflector.

In the FIRS construction, applied in the former USSR in 60-ies, the role of external cover was played by rubber-fabric material with binder, based on silicon polymers.

It is known multilayer system, consisted with intumescing cellulose material, impregnated by products of amine condensation with aliphatic aldehydes. Thermoplast with aluminum foil was used as protective layer. The following materials were recommended for internal thermal insulation: mineral and glass wool, vermiculit, asbestos paper [1].

Efficient and good for practice system has been suggested for protection of fuel tanks [2]. The scheme of layer positioning in the construction is shown at Figure 1. The method of heat shielding container producing, differing from others by layer of plastic foam between external cover and container, is described in FRG patent [3]. Construction and protection of container for material storage, suitable to overheating, and description of the case, protecting magnet carriers, are shown in works [4,5]. Here the presence of heat shield external and thermal isolating internal layer is foreseen.

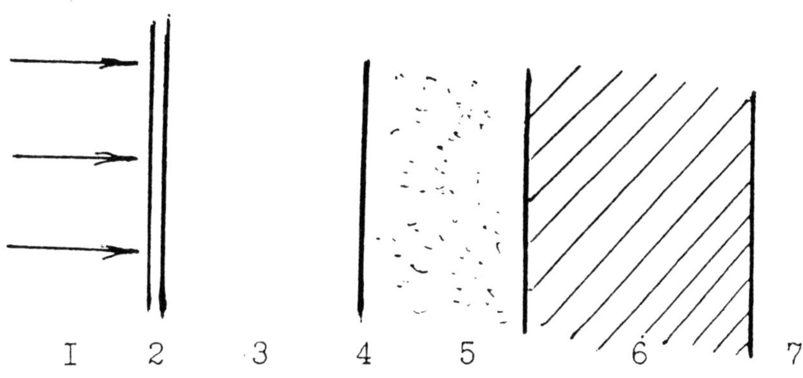



**Figure 1.** Scheme of layer disposition in construction of fire and heat shield of fuel tanks: 1 - flame; 2 - thin steel cover; 3 - vacuum or air; 4 - reflecting surface; 5 - thermal insulation; 6 - tank wall; 7 - fuel.

USA patent [6] describes FHSM and the method of its obtaining. The material consists with 2 layers: internal thermal isolation layer expands at heating, external layer plays the role of ablation cover. The latter swells up at heating, absorbing large amount of heat, and prevents flame influence on internal layer. Technology of FHSM obtaining includes the following operations: Thin material layer of pored and cellular structure is obtained by pressing. It is covered by the layer of thermoplast with low decomposition temperature (93-204, 4°C), and then the layer of intumescent coating.

All the above described fire and heat shield covers represent complex constructions, applying various methods of heat dissipation.

The questions, meaning understanding global chemical and physical processes, which control gasification of polymer material and heat transfer in the system, take central place in working out efficient FHSM.

Heat energy of external source, feeding the surface of the material, can increase its temperature so that material decomposition begins. Combustible products of decomposition, been formed, are able to ignite and provide combustion of the material itself and fire spreading. That is why the problem of efficient FHSM creation is shortly connected with that of the decrease of polymer material combustibility. The latter, in its turn, can be considered as inherent component of the first part.

In unusual conditions heat transfer proceeds from more heated (hot) bodies to lower heated (cold) ones by heat conduction or convection. In conditions of fire the main type of heat transfer from flame or heated body surface to FHSM is radiation. Radiative heat flow, incident on body's surface, is reflected, absorbed and passes through the body, i.e. is divided into three components: reflected, absorbed and passed through [7-9].

Radiative heat flow, passed through the cover, feeds surface of protecting sample. Absorbed heat flow is accumulated by the material, transforming into various sorts of energy [10,11].

Level and spectral characteristics of radiative flow, absorption properties of the material, reflection ability of its surface relative to emission spectrum of falling flow influences the amount of energy, absorbed by polymer material.

Reflection ability depends on angle of beam incidence, state of reflecting surface, wave length of incident flow, reflective indexes of the media. Beams with wave lengths much smaller than the size of surface irregularities disperse. Beams with wave lengths much longer than this size pass freely.

Transmission and absorption coefficients depend on chemical composition and structure of a substance. If absorption coefficient is small, radiative flow passes through all the material, and its heating is decelerated. However, if absorption coefficient is large, radiative energy is absorbed near the surface, and thin layer of the material is heated up rapidly to critical temperature of its decomposition. In this case maximal temperature in the material is obtained at its surface.

It is important from the point of heat shield to increase reflective ability of the material. However, polymers and other organic compounds have low reflection coefficients [10]. Reflective ability can be increased by surface modification as well as by introduction of fillers into the materials, able to reflect radiative energy in infrared spectrum range of radiative flow. Oxides of various metals can be applied as such fillers.

The depth of heated layer of the material and time of its heating up to critical temperature is defined by nature of polymer material, its heat-physical properties, amount of feeding heat [11]. The depth of surface layer and time of its heating decreases with the increase of heat flow intensity, feeding material surface, and the decrease of heat inertia ($\lambda\rho c$). Polymer materials with porous or cellular structure, having low heat inertia, possess higher thermal isolating characteristics. However, there appear the danger of their faster ignition and flame spreading along the surface of these materials in comparison with monolythic analogues.

Polymer FHSM are composite materials. They contain many components for various direct purposes (fillers, plastifiers, combustion decelerators, etc.). Physical and chemical processes of the transformation of components and the whole material, proceeding under the influence of absorbed heat energy, are complicated. The mechanism of transformation processes and their role in heat and mass transfer, their influence on efficiency of shielding action of the cover are specific and not well studied, yet. However, it is evident, that in the present case processes are preferable, accompanied by heat absorption and promoting its removal into the surroundings.

Phase transitions of substances from one state to another are corresponded to physical endothermal processes: crystalline substance melting, power-consuming processes of evaporation and sublimation, transition from solid to viscous-flowing state.

Chemical reactions of molecular bonds dissociation are accompanied by heat absorption. However, total heat effect of the process of substance decomposition depends on the direction of further elementary reactions and their energetics. In dependence on polymer nature reactions of thermal decomposition can pass by various ways.

Usually two large groups of polymers are separated, which at high temperature degradate practically full and from low-molecular volatile compounds or are changed by complicated structural transformations into carbonized products. Polymers of the first group, despite the difference in degradation mechanism (for example, reverse depolymerization or random break of the main chain with outlet of smaller molecular fragments), are characterized by endothermal effect. Polymer of the second group show tendency for exothermal reactions of side substituent cutting off with the formation of conjugated double bonds, cyclization, condensation, recombination and intermolecular cross-linking, which lead to carbonization of nonvolatile residue. The process of polymer decomposition of this group is accompanied at early stages by bond break in the main macromolecular chain and outlet of volatile low molecular compounds. That is why total heat effect of low-temperature stage of the degradation is often endothermal or close to neutral one as a result of compensation by the heat of exothermal reactions. More high-temperature stages of decomposition of coke-forming polymers are exothermal, as well as the reactions with oxygen participation.

Coke-forming thermoplastics at degradation during the first stage expand and swell up often, promoting the formation of foamed coke. Morphological structure of coke layer influences sufficient polymer inflammability, heat shield properties of the cover. On one hand, coke layer with cellular or porous structure has high absorption ability and accumulates large amount of heat energy. According to temperature increase of carbon layer, heat amount increases, radiated from its surface into the surroundings, as well as transferred to lower layers of the material [7]. On the other hand, foamed coke has sufficiently lower density, $\rho$, and heat conductivity coefficient, $\lambda$, heat capacity decreases, $c_p$, in comparison with initial polymer material. Thus, thermal isolating ability of foamed coke layer begins to play dominant protection role in FHSM action. Thus, coke strength is of important meaning. The change of cover volume at constant sizes of protecting surface, thermal contractions and gas product separation are the cause of force appearance in FHSM and partial degradation of polymer material. Cokes with large-cell structure are usually friable and inefficient as thermal isolators. Understanding physical and chemical processes, participating in the formation of strong foamed coke at cover working is necessary for the creation of new efficient FHSM.

Theoretical models of intumescent coke formation are shown in works [12,13]. Mathematical models of carbonizing polymer pyrolysis in conditions of unidimensional heating are represented by various investigations [14,15]. Paper [15] considers variants of one-stage and double-stage coke formation, the influence of surface heterogeneous chemical reaction is analyzed, as well as the formation of porous structure in consequence of appearance and motion of bubbles of volatile degradation products. Suggested models are based on the number of suppositions and simplifications. In particular, complex kinetics of pyrolysis is described by one global temperature dependence only, corresponded to the Arrhenius law. It is supposed that substance transformation into coke begins immediately after reaching some threshold temperature, specific for each polymer.

Heterogeneous reaction of coke oxidation (smouldering) leads to heat generation and accumulation inside the cover. In this connection polymers, forming nonsmouldering solid residue and inflammable gases during degradation, are preferable for FHSM creation.

The decrease of combustibility of polymer materials can be attained by physical and chemical methods [16]. The following methods, above mentioned already, being interesting in connection with

FIRS, should be attributed to physical methods of influence on the contribution process, directed to the decrease of heat feeding of the material, the increase of heat outlet into the surrounding, weakening of conditions of mass transfer to the zone of combustion reaction. The following chemical processes are practically interesting as influence measures: directed change of the composition and structure of a polymer, composition and component ratio in the material. This influences kinetics and mechanism of reactions of polymer material degradation, its ignition and combustion.

The following directions can be pointed out in working out flame retardant (polymers) [16]: the synthesis of inflammable or hardly combustible polymer; surface and voluminal chemical modification of polymer materials; application of flame retardants; purposeful polymer blending; introduction of inflammable fillers into the composition; applying fire shield coatings; combination of all the above-mentioned methods.

Resuming all of this resistivity of the material to ignition should increase. The rate of flame spreading along the surface and heat release rate should decrease. Beside these the most important properties of combustibility, fire safety of materials is characterized by quantity of toxic products of decomposition and combustion, the level of smoke formation.

Stable flame combustion of polymers includes the totality of mutually adjusted gas- and solid-phase processes. This adjustment is stipulated by level and rate of heat flow feeding from polymer flame to its surface. It can be expressed by concept of Spoulding mass transfer number, B, taking into account radiative component and heat losses or combustion efficiency [17]. Mass transfer number, B, is connected directly with gasification rate, $m''$, by the following expression:

$$m''=h\,ln(1+B)/c_p,$$

where h - convective coefficient of heat transfer; $c_p$ - heat capacity of the polymer material. In respect, mass transfer number, B, can be represented as follows:

$$B=[\{Y_{ox}\Delta H_c(1-\chi_R)X_A/r\}-c_p(T_s-T_o)]/\{L(1-E)\}$$

Here $Y_{ox}$ - mass concentration of oxygen in the surroundings; $\Delta H_c$ - the heat of full combustion; $X_A$ - combustion degree; $\chi_R$ - radiative energy part; $T_s$, $T_o$ - temperature of surface and medium, respectively; L - the heat of polymer material gasification; E - combustion efficiency. The latter, taking into account radiative heat losses from the surface, is expressed as follows:

$$E = (q''_e + q''_{fr} - q''_{rr})/m''L,$$

where $q''_e$ - external heat flow; $q''_{fr}$ - radiative reverse heat flow from fire; $q''_{rr}$ - radiative heat losses from the material surface. Other sorts of heat losses from the material can be calculated in analogue.

Polymers with low values of combustion heat and high values of gasification heat are relatively lower combustible materials. In this cause special interest is in inorganic polymers, the main chains of which consist with silicon, nitrogen, fluorine and other non-carbon elements, or thermostable polymers with aromatic or heterocycle structure [16]. However, the price of such polymers is quite high. Moreover, their processing into articles is often difficult, or hydrolytic stability is not enough. Another way of the decrease of the material combustibility are more accessible.

Thermal stability of a polymer can play an important role, reflectings its combustible characteristics. It was found, for example, that the existence of anomalous structure unit in polymethylmetacrylate chains decreases thermal stability of the polymer, leads to the increase of gasification rate by 25% and fire spreading rate along the material surface by 4 times, comparing with more stable sample [17]. However, such interconnection between thermal stability and polymer combustibility is not of universal character, because beside gasification rate the content of degradation products is important greatly. Flame retardant polymer materials are often characterized by more lower decomposition temperature, comparing with their analogues.

The most wide-spread approach to the decrease of polymer material combustibility is the application of flame retardants, performing various direct functions simultaneously (fillers, plastifiers, foaming agents, etc.).

Flame retardants represent substances, containing the following chemical elements: B, P, halogens, nitrogen, various metals. Mechanism of their action depends on nature of an element, chemical composition of a substance and polymer matrix, conditions of heat influence on the system. A number of works is devoted to this question [18, 19].

Combustion decelerators are separated as follows: substances with the mechanism of action in gas and condensed in dependence on the zone of leading influence.

Halogen-containing compositions are corresponded to typical flame retardants of gas-phase action. Volatile products, formed as a result of their decomposition, play the role of inert dilutors of flame medium and inhibitors of gas-phase chain reactions. Bromine containing substances are more efficient, than chlorine containing flame retardants. However, temperature influences sufficiently the efficiency of halogen-containing substances. In rigid fire conditions at temperatures of 1200-1300°C and higher bromine begins to show oxidative properties and to catalyze chain gas-phase reactions. Thus it performs unwanted influence on combustion process, on the whole [20].

Most of fluorine containing flame retardants are characterized by leading action mechanism in condensed phase. During polymer material combustion these substances are able to decompose, forming oxides and acid derivatives of fluorine. Dense surface layers of polyacidic compounds of fluorine create physical barrier from oxygen, reaching lower layers of the material, and diffusion of combustible products of decomposition into the zone of flame reaction.

Fluorine-containing compositions promote the increase of carbonized involatile residue yield at polymer decomposition. Thus there is observed the change of direction of coke oxidizing reactions to the side of smouldering process suppressing. The efficiency of fluorine containing flame retardants increase in the presence of nitrogen containing compounds.

Just as fluorine containing flame retardants, surface vitrious protective layers are able to form borates, silicates, various salts of alkaline metals [21].

It should be mentioned that the decrease of polymer material combustibility is performed mostly often according to mixed mechanism with participation of flame retardant in processes in condensed and gas phase. The part of one or another action mechanism depends on type of flame retardant and polymer nature.

A great number of substances was suggested as decelerators of combustion and smoke supressants [22]. Beside simple substances with high content of an element with flame extinguishing action (red fluorine, for example) some mineral and synthetic compositions were suggested: asbestos, talcum [23], bismuth and molybdenum compounds or their mixtures [24], oxide or hydroxide of aluminum [25], ammonium salts [26], carbides of metals [27].

In a number of works on the decrease of polymer material combustibility fillers are separated as active and inert ones. However, this separation is relative. It is necessary to take into account concrete temperature condition of material working, the presence of oxidant and other factors. Thus, asbestos fiber at temperatures below 900°C plays the role of usual inert dilutor with low heat conductivity for combustible part of polymer material. At the temperature of 900-1400°C there happens the separation of crystallizational water. Endothermal process of rebuilding chemical structure of asbestos happens at the temperature over 400°C. Chemical interaction between asbestos components and products of polymer degradation is possible. It is usual that action of inorganic fillers in polymer matrix at high-temperature heating is conducted to the decrease of concentration of combustible component, the change of thermophysical properties of the system, physical transformations of the filler. However, it could be seen a surprising coincidence: some inorganic compounds, used as flame retardants, are catalysts for reactions of polymerization and polycondensation at polymer synthesis. That is why it can be supposed that such compounds are able to perform catalytic functions according to products of decomposition during polymer combustion, to accelerate reactions of cross-linking and coke-formation.

Reactions of hydrocyclization of hydrocarbons and their aromatization in presence of catalysts show great interest. The following substances are active among these catalysts: $TiO_2$, $Co_2O_3$, $Al_2O_3$, aluminum phosphates [29]. Action mechanism of highly efficient flame retardants of intumescent type is connected with catalysis of coke formation reactions, proceeding with the participation of polymer or char-forming component, specially introduced into intumescent system [29].

The number of flame retardants, known till now, is great. However, definite demand, narrowing greatly probable choice, are put forward at working out concrete materials.

Main general demands for flame retardants of polymers are the following: compatibility with the polymer; stability at the temperature of composition or material processing into article; absence of negative influence on technical, physico-mechanical and other exploitational properties of the material; nontoxicity, corrosion inactivity. It is natural that the type of polymer basis plays definite role here.

The following factors were taken into account at polymer selection for FHSM: accessibility of raw basis, developed industrial production, relatively low price, decreased combustibility. Elastomers are preferable for creation of impact-resistant FHSM.

The analysis of the information existed leaded to the conclusion that halogenized polymers are mostly interesting as polymer based of FHSM.

Native industry produces various chlorine containing polymers: homopolymers and copolymers of vynilchloride and vynilidenchloride, chlorinated butylrubber, polyisoprene and polyvynilchloride, chlorinated and sulfochlorinated polyethylene, etc.

Additional chemical modification of polymers by their halogenization broadens spheres of production application [30, 31].

Polymer chlorination can be performed in medium of organic solvents, in water suspension, and in condensed phase by the influence of gas or liquid chlorine. Sulphuration can be performed simultaneously with chlorination. The process is initiated by the influence of visible or ultraviolent, or ionizing radiation, with the help of substances, forming free radicals during decomposition (organic peroxides and hydroperoxides, diazocompositions, metalorganic substances) [30].

Properties of chlorinated polymers depend on nature of initial polymer, chlorination degree, method and conditions of its performance [30]. Wide application in technics was found for chlorinated and sulfochlorinated polyolefins.

Let us consider features of production and their properties.

## 3. PROPERTIES OF CHLORINATED AND SULFOCHLORINATED POLYETHYLENES

Modification of polyolefins by halogenization shows a number of features, stipulated by chemical inertia of initial polymers, the presence of amorphous and crystalline area in their structure. Reactions of halogenization can be accompanied by degradation and cross-linking of the main chains of macromolecules [32, 33]. The process is preferably performed in solvent medium or in suspension in water.

At room temperature polyethylene is not soluble by common solvents. In dependence on crystallinity degree, PE begins to solve in chlorinated alyphatic and aromatic hydrocarbons at temperatures over 53°C. PEHD is solved fully in $CCl_4$ at the temperature, close to its boiling. PEHD solubility is observed at higher temperatures (80-100°C) [30]. Solubility of chlorinated PE depends on chlorine content in polymer: At chlorine content up to 30 mass % solubility in organic solvents increases, then it decreases, and at 50-60 mass % of chlorine polymer becomes insoluble [30].

USA patent describes the method of PE chlorination in solution in $CCl_4$ in presence of radical initiators [34]. Application of chlorobenzene, possessing higher solving ability, allows to increase the reaction rate [35]. Mixtures [38], as well as initiators of radical type are applied for chlorination: dinitrilisobuyric acid and peroxide of benzoyl or lauryl, dihydroxyperoxide of chlorate together with acetylcyclohexylsulfonyl peroxide. ICI Company (Great Britain) for the first time performed PEHD chlorination in suspension in $CCl_4$ or acetic acid in presence of metal chlorides [30].

Liquid chlorine has been suggested as chlorinated agent [39], but gas reagent is applied often. The product, containing up to 40-44 mass % of chlorine [40], has been obtained at PE chlorination in suspension in $CCl_4$ by gas chlorine. Water suspension method of PELD chlorination suggests the application of surface-active substances [41].

Chlorinated polyethylenes, produced by companies of different countries, with chlorine content of 25-50% represent amorphous or low crystal materials with the tensile strength from 8 up to 17 MPa and relative elongation at break of 350-800%. At wider range of chlorination (from 14 to 70%) various materials are obtained: from thermoplastic polymers to friable vitrious products [42]. As a result of modification the structure of final polymer changes. At chlorination or sulphochlorination in solution crystallinity breaks stronger, than at halogenation in suspension. Polymer crystallinity is kept partially in the last case at chlorine content lower 55 mass % [32].

The investigation of structure of chlorinated PE has shown that $-CCl_2-$ groups are absent in polymer structure. At the chlorine content lower 55% vycinal groups -CHCl-CHCl- are absent also [32]. First of all, at chlorination amorphous spheres of polymer take part in the reaction, and then more regulated spheres are implicated into it. The reactions of hydrogen atom substitution by chlorine proceed statistically, leading to spontaneous distribution of chlorine in polymer chain. Submolecular structure of the polymer changes from spiral for PE to spherolytic (monocrystal) at chlorine content of 8%. At the increase of chlorine content up to 14-18 mass % short fibrils are observed, and at the increase up to 22-30 mass % of chlorine - packs and globules appear [32]. Full amorphization of polymer appears at chlorine content over 55%.

Physical state of PE according to chlorination in solution changes as follows: [41]:

| Polymer State | Chlorine Content, mass % |
|---|---|
| Semielastic thermoplast | 10 |
| Thermophastic elastomer | 20 |
| Thermoplastic, flexible, viscous elastomer | 30 |
| Thermoplatic elastomer | 40 |
| Semielastic thermoplast | 50 |
| Rigid thermoplast | 60 |
| Friable polymer | 75 |

Glass formation temperature of polymers at chlorine content of 25-35 mass % lies in the range from -20 up to +30°C, and at the content of 28-73 mass % of chlorine increases up to 100-180°C. The change of softening point is of parabolic character with the minimum at 35-40 mass % of chlorine [30]. Chlorinated PE are stable for influence of acids, weak alkali, salt solutions, show decreased combustibility. Under flame influence chlorinated PE does not melt, but is carbonized. The outlet of coke residue increases with the increase of chlorine content [31].

Numerous works [41-45] are devoted to the investigation of chlorinated PE thermal degradation. Comparing with initial polymer, the mechanism of chlorinated pE degradation becomes more complicated. The process becomes a multistage one, including chain degradation as well as reactions of substituents separation with the formation of polyene fragments and their further transformation into cross-linked aromatized structures.

Caoutchouc-like products of polyethylene sulphochlorination (PESC) are greatly interesting for FHSM working out. They are obtained by bubbling mixture of gas chlorine with sulfur anhydride (in molar ratio 2:1, preferably) through solution of PE in $CCl_4$ at 70-75°C in present of catalytic amounts of peroxides or asobisisobutironitrile. Native industry produces 6 types of sulfochlorinated polyethylene, based on PEHD [46], for preparing various articles. Foreign assortment of this production is more different. Thus, DuPont Company, USA, creates production of 8 types of PESC, based on PEHD as well as on PELD.

PESC, produced by native industry, containing 26-36 mass % of chlorine and 0.85-2 mass % of sulfur has the temperature of the beginning of decomposition 115-150°C, tensile strength is about 15.0 MPa, relative elongation at break 300-350% [42]. Physico-mechanical properties change at broadening of chlorination limit up to 59% of Cl and sulfur up to 4,4 mass % [30, 49].

It was found in work [47], devoted to the study of influence of structural factors on PE modification, that polymer branching does not influence sufficient properties of final product. More important role is played by molecular mass and MMP of initial polymer. Detailed analysis of chlorinization degree influence on relaxational properties and phase state of PESC is shown in work [49]. PESC with chlorine content lower 59% and sulfur lower 4.4% have been obtained on the basis of polyethylene of high density ($\rho$=966 kg/m$^3$). Spectrum of relaxational and phase transitions, stipulated by various types of molecular motions, was exposed by methods of dynamic mechanical spectrometry, linear dilatometry and DSC. Crystallinity degree of polymer decreases at the increase of chlorine content. Full amorphisation of PESC is observed at chlorination degree of 30 ± 2 mass %. Shear modulus decrease is a consequence the decrease of polymer crystallinity degree (Figure 2, curve 1). Shear modulus was measured at temperature, corresponded to low-temperature boarder of highly elastic state of amorphous part of polymer. Folding is observed at curves of the change of density and temperature of PESC vitrification (Figure 2, curves 2 and 3) at critical chlorination degree, corresponding to full amorphization of polymer. DSC data prove the conclusion about the change of crystallinity of polymer at PE sulphochlorination. Endopeak on thermograms of PESC, stipulated polymer crystal phase melting, changes its form, shifts to the lower temperature, decreases by value with the increase of chlorine content up to 30 mass %.

                          *M.A. Shashkina et al.*

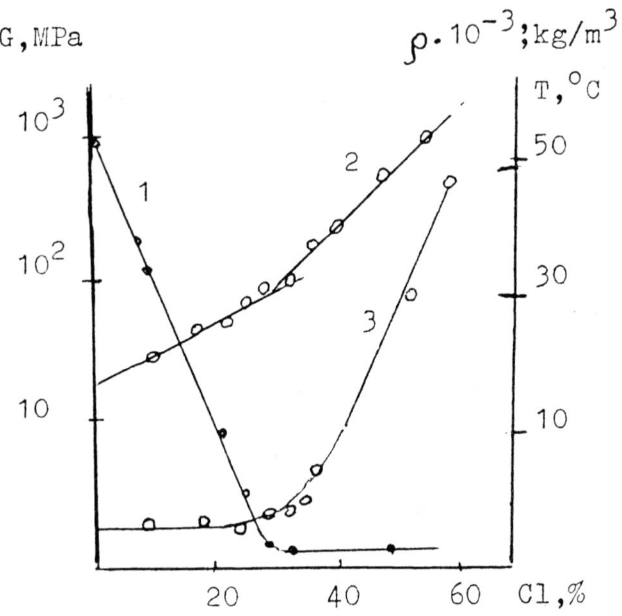


**Figure 2.** The influence of chlorine content in PESC on the change of shear modulus (1), density (2) and glass formation temperature (3) of a polymer.

In comparison with chlorinated PE sulfochlorinated product possesses important advantage: better ability for vulcanization. Vulcanization is performed, first of all, by sulfochlorinated groups. To obtain impact-resistant vulcanizate optimum sulfur content in polymer is ~1.5%. PESC differ by high light-, ozone-, oil- and petrol resistance, decreased combustibility [50]. Non-vulcanized PESC are stored for a year without changing their properties [51]. PESC vulcanizates are extremely resistant to ozone aging. This index relates them to the group of extremely stable polymer materials. PESC vulcanizates attract attention by high resistance to various atmospheric influences, ionizing radiations [30]. In tropic conditions they retain their properties for a long time, endure the influence of mold and microorganisms.

High static strength of PESC vulcanizates is observed even in absence of strengthening fillers. This friability temperature varies in the range from -55° up to 60°C. Vulcanizates possess satisfactory resistance to tearing and attrition, retaining at increased temperature. They are characterized by high strength to numerous elongating and bending loads, to decrepitation and heat aging. Maximum working temperatures for the most number of articles from PESC fall within the range of 130-160°C. At 100°C working time of PESC rubbers is measured by years. Properties of vulcanizates depend sufficiently on conditions of vulcanizing agent [52,53].

Vulcanization of elastomers represents complex process of linear macromolecule transformation into spatial-net structures with relatively rare positioning of cross bonds. As a result product loses solubility and thermoplastic properties, acquires high elasticity, strength and a number of new valuable properties. Various reactionally able atoms and groups of PESC can participate in the formation of spatially-net structure: chlorine-sulfone groups, chlorine and hydrogen atoms, unsaturated double bonds, formed at polymer dehydration [54, 55]. Chlorine atoms, present in $\beta$-position to chlorosulfonic groups, are mostly labile and possess the ability for catalytic dehydrochlorination.

The presence of various reactionally able atoms and groups in PESC allows to apply various vulcanizing agents and systems, organic and inorganic compounds for intermolecular cross-linking of chains. The following substances were suggested for this purpose: metal oxides, sulfur and sulfur-containing compounds, peroxides, dissocyanates, nitrogen-containing bifunctional compounds, polyatomic alcohols, etc. [54-58]. Mostly efficient are PESC vulcanizing systems, included oxides of lead and magnesium [54]. At present combined vulcanizing systems, included oxide or salt of polyvalent metal (10-50 mass parts), organic acid (2-10 mass parts), vulcanization accelerator - sulfur containing compounds (0.5-10 mass parts) are widely used [30].

It has been found that real cross-linking agents in these systems are accelerators of sulfuric vulcanization. It is supposed [30], that metal oxides make no direct influence on chemical transformation of PESC macromolecules. They participate in the formation of spatial-net structure as sorptional surface, dispergator of real cross-linking agent and absorbant of gas products. According to another conception, metal oxides promote chlorosulfonic group transformation into more polar ones of basic salt type [58]. At the initial stage in presence of water and activators of acidic type (fatty acids, for example) or vulcanization accelerators (thiurams, for example) at polymer heating chlorosulfonic groups hydrolize, forming HCl and -SO$_2$H groups. The latter participate in the formation of metal-sulfonate vulcanizational bonds. Accelerators of sulfuric vulcanization react with chlorosulfonic groups. There appear cross linkages or side substitutors. The latter are able to associate with each other and with polargroups on the surface of metal oxide, also. The bonds of polymer-metal oxide type, formed as a result of adsorption or hemosorption, are stable at usual temperatures of material exploitation.

The following substances are used as sulfur-containing accelerators of vulcanization, participating in formation of spatial-net structure: thiazoles, thiurams, dithiodicarbamates. Aldehydamine accelerators call scorching of PESC during its processing [54, 55]. Guanidines are used only as secondary accelerators in cooperation with thiazoles [30]. High vulcanizing activity of PESC is shown by salts of hexamethylenediamine and diatomic acids (hexamethylenediammonium sebacinate, for example [57]) or products of their condensation [59]. In this case, magnesium oxide increases cross-linking degree, and sulfur additives accelerate vulcanization [57]. Dithiomorpholine is recommended, as promoting high concentration of cross bonds in PESC vulcanizate in absence of additives, and at presence of magnesium oxide - high strength properties of vulcanizate [60]. Combination of magnesium sulfide with diphenylguanidine and 2-mercaptoimidasoline [61, 62] is very effective. Polyatomic alcohols (pentaerythritol, for example [63]), chlorinated aromatic compounds (para-xylene hexachloride, for example [64]), find the application.

The regimes of PESC vulcanization at the application of various vulcanizing systems are shown in [65]. There are the description of composition formulations, based on PESC for obtaining various materials of different application, possessing increased stability to heat and light-ozone aging, influence of aggressive media and water [66].

Technology of PESC processing does not differ from that for elastomers, applied for a long time in industry. Thus, the equipment, usual for rubber industry, is suitable. It should be mentioned that nonvulcanized PESC elastomer is more thermoplastic, than natural or many other synthetic caoutchoucs. That is why it needs no preliminary plasticization [30, 48]. Preparation of mixtures on rollers or in rubber mixer is accompanied by sufficient heat outlet, that can lead to subvulcanization of compositions. To avoid this process it is necessary to make mixing as quick as possible.

Mixtures, based on PESC, are formed satisfactorily by the methods of pressing, extrusion, calendaring and pressure molding. Mixtures, based on PESC, should be heated before extrusion or calendaring because of their increased viscosity. Vulcanization of mixtures, based on PESC, is performed at temperature of 120-150°C. Higher temperatures of processing can lead to the appearance of porosity, shells and other defects on the surface of the material. Vulcanization of PESC compositions is performed in press or by sharp vapor at pressure up to 1.8 $MN/m^2$. PESC is characterized by wide plato of vulcanization, and mixtures on its basis are stable for subvulcanization.

# 4. FIRE AND HEAT SHIELD MATERIALS, BASED ON SULFOCHLORINATED POLYETHYLENE

It is known the application of chlorinated and sulfochlorinated PE itself or in mixtures with caoutchoucs and thermoplastics for working out materials of decreased combustibility. For example, there were described impact-resistant compositions of decreased combustibility, based on chlorinated and sulfochlorinated PE with chlorine content of 25-45 mass %[67].

There have been suggested compositions and compounds for obtaining shield coatings of decreased combustibility, included additionally halogen-containing flame retardants synergists and inorganic fillers [68-73]. PESC vulcanizates with low chlorine content are characterized by decreased combustibility, but resign in this property vulcanizates, based on chloroprene caoutchoucs. New types of PESC with increased chlorine content are equal to chloroprene or exceed it by flame influence resistivity.

The analysis of information about properties of chlorinated and sulfochlorinated PE leaded to the conclusion that for working out impact-resistant FHSM for FIRS, it is mostly expedient to use as polymer matrix an industrial product, obtained by PEHD sulphochlorination, with chlorine content of 27 mass % and sulfur content of 1.3-1.5 mass %.

To obtain compositions with high technological characteristics, stable to subvulcanization, combined vulcanizing system has been used, included magnesium oxide, sulfur vulcanization accelerator and some other directed components. Silicon and antimony oxides, chloroparaffin of CP-470 mark with chlorine content of 47 mass % were introduced into the composition.

Silicon oxide (white soot) is an active filler for PESC, allowing to obtain rubbers with high physico-mechanical properties and, in particular, with high resistence to tearing and numerous bending loads. Chloroparaffin, efficient plastificator for non-coloured PESC rubbers, plays simultaneously the role of flame retardant. As it is known, antimony oxide is synergyst for halogen-containing compositions, increasing fire shielding effect.

Compositions were obtained by mixing ingredients on rollers. Samples of rubbers were prepared by the formation and vulcanization of compositions in press at temperature of 140°C and pressure of 5 kg/cm$^2$.

Optimal formulation was found for a composition as a result of verification of composition and content ratio, study of the influence of these factors on the final exploitational properties of FHSM, mentioned below, is the optimium between polymer and vulcanizing system: 100:27.5 mass parts.

74          *M.A. Shashkina et al.*

To understand mechanism of fire shielding action of the cover for FIRS, the influence of applied metal oxides and chloroparaffin on combustible and thermal properties of PESC vulcanizates was studied in detail.

Table 1 shows the results, illustrating the influence of nature and concentration of metal oxide in PESC vulcanizate on the value of oxygen index, OI, and the rate of flame spreading along sample surface. Oxygen index was measured by standard method GOST 12.1. 044-84, and at the sample combustion in horizontal and vertical upward direction. In these cases sizes of the samples and reactor, the rate of flow of nitrogen mixture with oxygen were corresponded to standard conditions of OI determination.

**Table 1.** The influence of metal oxides on flammability of PESC vulcanizates

| Indices | Direction of flame spread | Metal oxide content, pph of PESC | | | | | | |
| --- | --- | --- | --- | --- | --- | --- | --- | --- |
| | | $Sb_2O_3$ | | | | $SiO_2$ | | |
| | | 0 | 30 | 50 | 70 | 30 | 50 | 70 |
| LOI, % | ↓ | 27,8 | 37,8 | 40 | 42,7 | 34,8 | 39,5 | 42,5 |
| $V_{fs} \cdot 10$, mm/s | ↓ | 2,01/28 | 0,56/38 | 0,52/40 | 0,85/43 | 0,7/36 | 1,18/40 | 0,73/43 |
| OI , % | ↑ | 20,5 | 34,5 | 35,8 | 37,8 | 27,5 | 29,5 | 30,5 |
| OI , % | <-- | 26,8 | 38 | 40,5 | 42 | 33,8 | 38 | 40 |
| $V_{fs} \cdot 10$, mm/s | without asbest | 0,66/27 | 0,5/38 | 0,7/41 | 0,53/43 | 0,90/34 | 0,84/38 | 0,45/40 |
| OI , % | <-- | 31 | 43,5 | 45,5 | 48 | 41 | 48,5 | 52,8 |
| $V_{fs} \cdot 10$, mm/s | with asbest | 1,64/31 | | 1,82/46 | 0,68/48 | 1,75/41 | 2,5/49 | 1,25/53 |

denominator-oxygen content in the atmosphere

The influence of thermal isolating support on limited oxygen concentrations and the rate of flame spreading was estimated.

OI of non-vulcanized PESC equals 22.5% . As it is seen from Table 1, the orientation of samples, direction of flame spreading, presence of support influences combustibility indexes. Sample thickness was 4-5 mm at testing. The influence of sample thickness on OI value is shown in Table 2 for vulcanizates of PESC of optimum formulation. At thickness of 3.5-7.3 mm OI is practically constant. The decrease of OI at thickness lower than 3.2-3.5 mm is stipulated by relative decrease of heat losses from flame into the surroundings.

**Table 2.** The influence of specimen thickness on oxygen index of FHSM with optimal formulation.

| Thickness, mm | Oxygen index, % | | | |
| --- | --- | --- | --- | --- |
| | Direction of flame spread | | | |
| | V | ! | without asbest | with asbest |
| 1,3 | 41,8 | 31,5 | 41,5 | 43,0 |
| 3,2 | 44,0 | 40,5 | 44,7 | 58,0 |
| 3,5 | 46,5 | >42$^X$ | 47,0 | 57,5 |
| 7,3 | 47,0 | >42$^X$ | – | 57,0 |
| 8,5 | 46,5 | >42$^X$ | 48,0 | 52,0 |

The values of OI at candle burning and at horizontal flame spreading along the surface of samples without support are close. At the support presence it is observed sufficient increase of OI and change of the rate of flame spreading (Table 1). This testifies that sufficient heat outlet from combustion zone through condensed phase and support happens. Apparently, support thickness is not enough for full heat isolation of the sample from fixer material. The rate of flame spreading along the sample surface increases with the increase of oxygen concentration in the surroundings. The most sufficient change of OI happens at transferring from candle combustion to upwards combustion. Thus, the rate of flame spreading is changeable, it increases with the flame spreading. In these rigid conditions of combustion vulcanized PESC without fillers is able to ignite and combust in air (OI=20.5).

At the introduction of metal oxides combustibility of PESC vulcanizates decreases. At filler content up to 30 mass parts antimony oxide is more efficient, than silicon oxide. However, the influence of metal nature is being leveled with the increase of metal oxide content up to 50-70 mass parts, if combustibility is determined in standard conditions. In this case the rate of flame spreading at combustion limit of filled vulcanizates of PESC is 2-4 times lower than for non-filled ones. High efficiency of antimony oxide in comparison with white soot is reflected in upwards combustion (Table 1).

It is interesting to point out that additional introduction of chloroparaffin into compositions, highly filled by antimony or silicon oxide, increases their technological properties, but it influences weakly

OI values of PESC vulcanizates. However, positive effect of chloroparaffin in OI increase was observed at the combination of these fillers.

PESC vulcanizates of optimum formulation, included all the abovementioned components, possess the highest OI values (Table 2) and the lowest rates of flame spreading along the surface in comparable conditions.

Fire shielding properties of materials for FIRS were estimated by the method of heat blow on the device of one-side influence of the heater (Figure 3). Planar heater with temperature of 1100°C is contacted with a sample surface of 80*80 mm size and 10 mm thick. With the help of thermocouple, which junction was situated at the shielded surface in the center of back side of the sample, recorded its temperature automatically. Estimating creatia were the rate of temperature change of shielding surface and its maximal temperature in the definite time moment. The methodics used reproduces simplified model of the process of fire shielding cover work. It allows to perform screening and fast selection of the most efficient FHSM.

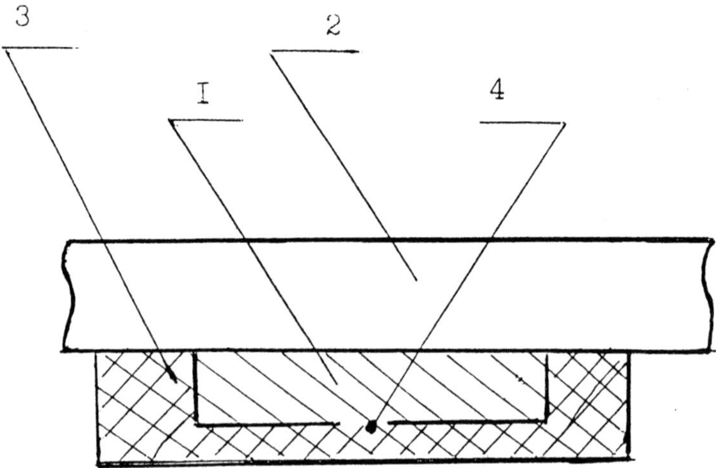


**Figure 3.** The device scheme for determination of fire and heat shield properties of materials: 1 - sample 2 - heater; 3 - thermal insulation; 4 - thermocouple juncture.

In conditions of testing fire shield properties of the cover, fast ignition of the sample happens. Flame spread along all the surface, contacted with the heater. Surface burning out is accompanied by polymer material carbonization. Carbonization front is moving inside the sample. White powder of mineral filler and then the layer of unchanged material were found under carbonized layer. Reaching 450°C at shielding surface, the material is carbonizing along all the thickness.

It was found with the help of X-ray pattern analysis, that there is practically no antimony in the carbonized product, and all other metals (Mg and Si) are safe as oxides.

The tests of fire shield properties of PESC have shown, that silicon oxide allows to decelerate heating rate of shielding surface sufficiently, than antimony oxide. However, additional introduction of chloroparaffin into the composition with $Sb_2O_3$ influences positively the decrease of temperature growth of the protecting surface (Figure 4). At the combination of two metal oxides and chloroparaffin the

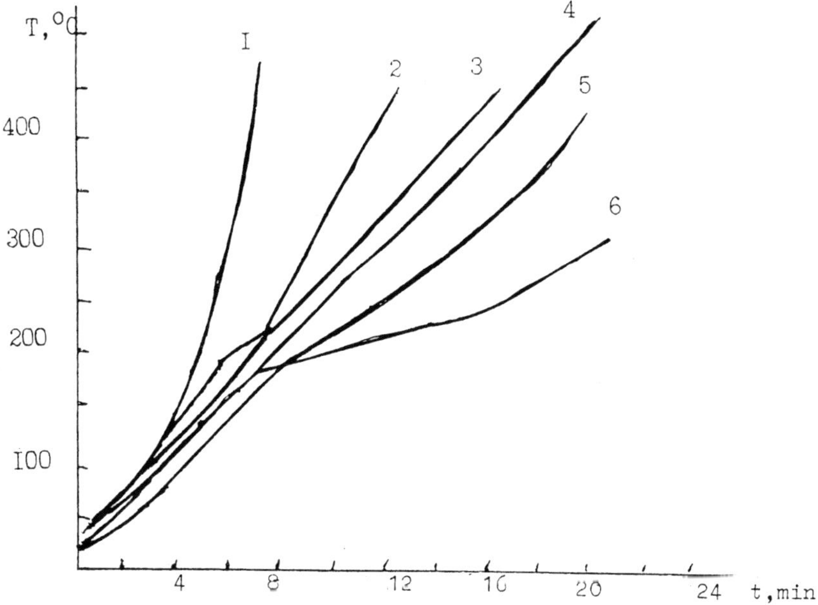

**Figure 4.** Fire and heat shield properties of compositions, based on PESC: 1 - vulcanized PESC; 2 - PESC with 50 mass parts of $Sb_2O_3$; 3 - PESC with 50 mass parts of $SiO_2$; 4 - PESC with 50 mass parts of $Sb_2O_3$ and CP-470; 5 - optimal formulation; 6 - composition with chloroparaffin.

efficiency of fire shielding action of the cover increases. The substitution of chloroparaffin by chlorobromoparaffin with close total content of halogens and Br=25 mass % gives the possibility more sufficiently decrease the rate of temperature increase of shielding surface (Figure 4, curve 6). However, this effect appears at test duration, exceeding 6-8 min. Apparently, at lower duration of the test there is not enough time for the formation of a layer of enough thickness, in that polymer transformation and the formation of foamed coke appears. At this stage, heat shielding of the surface is sufficiently determined by thermophyscial properties of the initial material, and its thermal activity.

The mechanism of heat shield action of the cover is complex. Thermal properties of PESC were studied, and heat flows from flame to the FHSM sample surface were estimated in order to clear up the sense of physical and chemical processes of that mechanism [74].

It has been found, that thermooxidative decomposition of PESC and vulcanizates, based on it, represent multistage process under dynamic conditions of heating (7 grad/min).

Metal oxide nature influences the range of rate of volatiles separation, as well as on coke residue outlet. Initial polymer possesses transition to highly elastic state at temperature over 100°C. Endopeak on thermogram of the sample with the maximum at 125°C is corresponded to this transition. Insufficient mass losses below 200°C (about 4 mass %) are stipulated by low molecular substances outlet, presented in the polymer. Intensive PESC decomposing is observed in the range of 230-400°C. The process proceeds according to the first order reaction with the rate constant of

$$k=4.3 \cdot 10^8 \exp(-94700 \ J/RT), \ min^{-1}$$

Splitting off of substituents in the chain and decomposition of chlorinesulfonic groups with the formation of sulfur dioxide and chlorinated hydrogen proceeds during this stage.

Vulcanized PESC decomposes intensively at temperatures over 182°C. Mass losses at heating up to 182°C do not exceed 2%. The first stage of decomposition in the range of 182-360°C is practically thermoneutral one unlike high temperature exothermal stages. It should be pointed out that despite the decrease of temperature of PESC decomposition beginning after vulcanization, its rate at the first stage is decelerated. Activation energy of vulcanizate decomposition increases up to 100 kJ/mole, and reaction order become the second one (Table 3). It is connected with conformational and spatial changes in net polymer structure, the decrease of macromolecules mobility as a result of vulcanization. The sufficient influence on the change of macromolecule mobility is caused by magnesium oxide, included into vulcanizing system.

**Table 3.** Kinetic parameters of thermooxidative decomposition for PESC compositions.

| Sample | Temperature range, °C | $E_{эф} \pm 2$, kJ/mol | $Z$, min$^{-1}$ | $n$ |
|---|---|---|---|---|
| PESC | 230-400 | 94,7 | $4,3.10^8$ | 1 |
| PESC (vulcanizate 1) | 182-362 | 110,0 | $1,6.10^{11}$ | 2 |
| I+SiO$_2$ (50) | 195-380 | 130,0 | $12.10^{13}$ | 2 |
| I+Sb$_2$O$_3$ (50) | 220-400 | 167,2 | $8,1.10^{17}$ | 2 |
| I+Sb$_2$O$_3$+CP-470 | 110-385 | 145,0 | $7,2.10^{14}$ | 1,5 |
| I+Sb$_2$O$_3$+SiO$_2$+SP-470 | 120-400 | 140,0 | $2,8.10^{13}$ | 2 |

Additional introduction of fillers (oxides of silicon or antimony) causes the increase of PESC vulcanizate stability. The opposite effect is observed at testing plastifiers - chloroparaffin or other compounds (Table 3).

As it is known, the value of efficient activation energy of polymer decomposition depends sufficiently on initiation reaction. The data obtained point out different character of molecular interaction of polymeric matrix with surface atoms or groups of atoms of metal oxides. According to the authors' opinion this interaction causes inactivation of labile PESC centres, initiating degradation of cross-linked polymer. Plastifier, promotes polymer degradation by the deterioration of the interaction with the surface of metal oxides.

Thus, it should be taken into account, that chloroparaffin and antimony oxide are reactionally able compounds. Chloroparaffin CF-470 decomposes at temperatures over 130°C during heating in dynamic conditions. Its OI is 25% [75].

Antimony oxide interaction with hydrogen chloride (the product of PESC and CP-470 dehydrochlorination) leads to the formation of oxychlorides and volatile antimony chloride [76,77]. The initial intermediate product - antimony oxychloride SbOCl - is instable and decomposes at 240-280°C according to the following reaction:

$$5SbOCl \rightarrow Sb_4O_5Cl_{(s)} + SbCl_3 \text{ (g)}$$

Further decomposition of $Sb_4O_5Cl_2$ at higher temperatures (400-570°C) limits the outlet of $SbCl_3$ into gas phase [76-78].

Similar antimony oxybromides are formed in the presence of bromine-containing compounds and $Sb_2O_3$. Compound $Sb_4O_5Br_2$ was found in the condensed phase, in combustion products of polyolefin compositions with synergetic mixtures with $Sb_2O_3$ and bromine-containing flame retardant [79].

The afficiency of decelerators action of this type depends on the ratio Sb:Hal in a composition, interaction character between $Sb_2O_3$ and halogen-containing compound, polymer nature. All these factors influence the amount of SbHal$_3$, coming to gas phase. This SbHal is the efficient reaction inhibitor in flame.

Characteristic feature of thermooxidative degradation of PESC compositions is the proceeding of secondary reactions with the formation of carbonized product and oxidation of the latter at temperatures over 500°C. Polymer transformation into carbonized structure, included fragments of condensed aromatic cycles, proceeds after dehydrochlorination and the appearance of insatured $\pi$-conjugated carbonic bonds in macromolecules.

Magnesium oxide and silicon oxide do not practically influence the outlet of coke residue. Taking into account moving off antimony oxide from the system as volatile derivatives, one can conclude, that in comparison with initial PESC vulcanizate in this case coke residue outlet increases.

It is interesting to look for the process, how atomic ratio of chlorine and antimony in compositions influences indexes of PESC rubbers combustibility. Compositions, containing just antimony oxide, possess Cl:Sb ratio change from 7.46 (30 mass parts of Sb$_2$O$_3$) down to 1.6 (70 mass parts). At additional introduction of chloroparaffin into highly filled system this ratio increases up to 2.16, coming close to the ratio Sb:Cl=2.26 for PESC compositions with 50 mass parts of Sb$_2$O$_3$. It is known, that the most large effect in the decrease of polymer combustibility with the help of synergetic mixtures of halogen-containing compositions and antimony oxide is reached at Cl:Sb ratio, close to optimum one for the formation of SbCl$_3$. There is no admiration, that in comparison with compositions, filled by metal oxides, the highest influence of oxygen index is observed at SbO$_3$ content of 30 mass parts. Additional introduction of chloroparaffin into highly filled composition does not provide optimal ratio of Cl:Sb=3 and does not lead to the increase of its OI. At the same time, the ratio Sb:Cl comes close to optimal one (3.04) for the formation of SbCl$_3$ at the application of chloroparaffin in the composition with 50 mass parts of Sb$_2$O$_3$. This influences the increase of the effect of fire shielding action of FhSM (Figure 4, curves 2 and 4). The mostly sufficient fire shielding effect is observed at the substitution of chloroparaffin by chlorobromoparaffin in the composition (Figure 4, curves 5 and 6), although atomic ratio Hal:Sb changes from 3:1 to 2:1 in this case.

The analysis of these derivatograms for FHSM shows, that the differences between them are useful just as low - temperature stages of decomposition. In the case of FHSM with chlorobromoparaffin slow decrease of the sample mass begins at the temperature over 130°C. And at temperatures up to 220°C, the beginning of intensive decomposition, mass losses are 6% (instead of 2% for the compositions with CP-470).

Additional stage with neutral heat effect and insufficient total mass losses is observed in the temperature range of 330-403°C. The outlet of involatile residue is similar at 490° and 535°C (Figure 5).

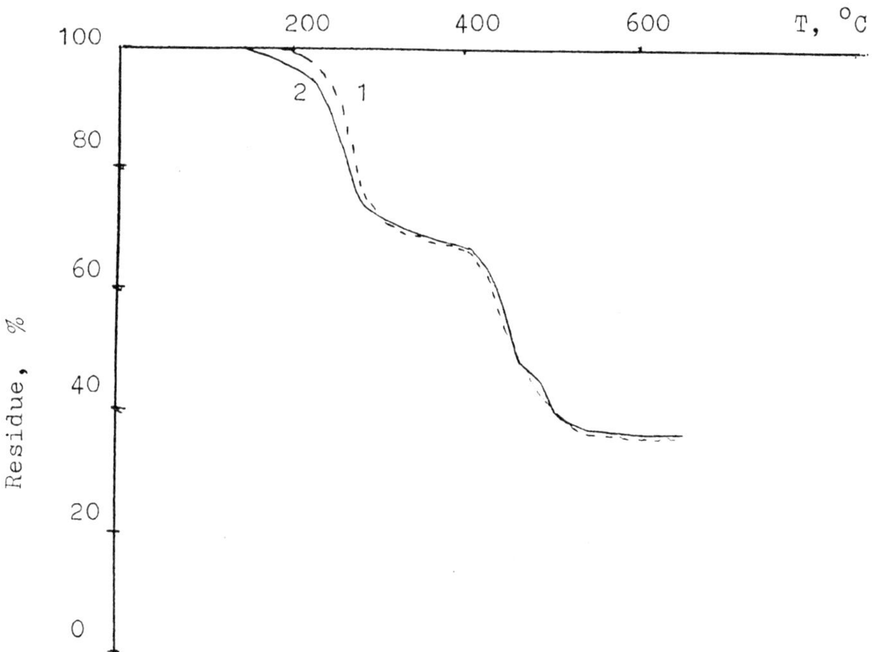


**Figure 5.** Thermal gravimetric curves for FHSM decomposition, containing CP-470 (1) and chlorobromoparaffin (2), at air heating with the rate of 7 degree/min.

The results obtained allowed to make a conclusion, that the following factors influence sufficiently the efficiency of FHSM: reactions in condensed phase, composition and amount of flame retardants, transmitting into gas phase. At the over fill of halogen in relation to antimony HHal and $SbHal_3$ appears in the gas phase. Nonreacted antimony oxide plays the role of inert dilutor in the condensed phase of combustible component of PESC vulcanizates. It is known, that $SbHal_3$ is more efficient inhibitor of flame reactions than HHal. The latter works usually as inert dilutor in the gas phase.

Antimony halogens ($SbCl_3$ and $SbBr_3$) possess double function in flame: i) they are the source of halogen-hydrogen; ii) they form antimony monoxide SbO, participating in the catalysis of recombination reactions of active radicals H, O, OH in flame by the formation of intermediate particles, such as SbOH [80].

Spatial zone and the time of presence in flame of inhibitor particles, influencing combustion process, increases due to different boiling temperature of antimony halogenides (223°C for $SbCl_3$ and 288°C for $SbBr_3$).

To determine the mechanism of flame spreading along FHSM surface temperature distribution in the combustion wave at the reverse flow of oxidant has been measured. Oxygen concentration in nitrogen-oxygen medium was found close to the limit of flame extinction (46%). The details of measurement and calculation of heat flows are shown in paper [74].

The following heat-physical characteristics of PESC composition were applied at the treatment of experimental results:

i) heat conductivity coefficient $\lambda$=0.33 W/m•K;

ii) specific heat capacity $c_p$=1.3 kJ/kg•K;

iii) Thermal diffisuvity $a$=1.6•10-7 m2/s;

iv) density $\rho$=1550 kg/m$^3$.

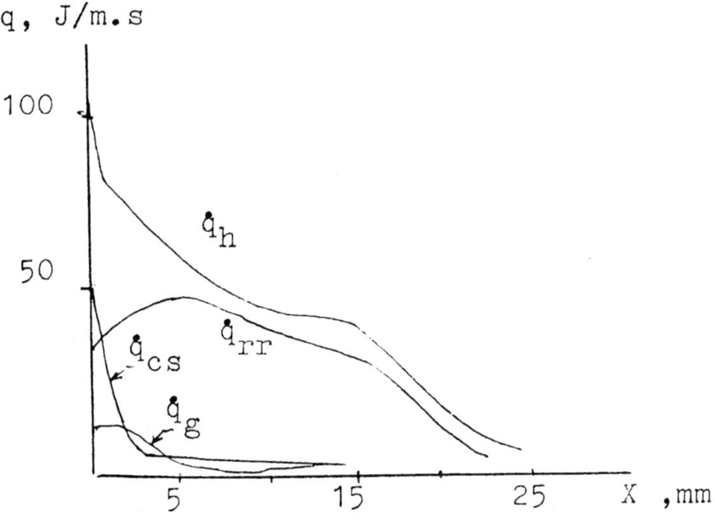


**Figure 6.** Distribution of heat flows during downward flame spreading on vertical surface of PESC vulcanizate sample, containing 30 mass parts of $Sb_2O_3$.

Temperature reaches 830°C in the flame edge at the distance of 3 mm from the sample surface. Maximal temperature of the flame outside the edge is 1230°C. Temperature at the surface beneath the flame edge is 620°C. However, in a far distance from the flame edge, the temperature on the surface increases up to 960°C, as a result of coke-formation and coke oxidation.

The flame edge was determined by the position of maximal temperature gradient in the gas phase. Figure 6 shows the distribution of heat flows on the FHSM surface.

Heat balance at combustion has taken into account heat flows, transmitted to the surface from flame by convection through the gas phase ($q''_g$), by heat conductivity through condensed phase ($q''_{cs}$), and by radiation ($q''_{fr}$), also. Heat losses are probable, because of irradiation and reflection of heat energy of FHSM surface:

$$q''_l = q''_{rr} + q''_{ref}$$

Thus, heat balance can be presented by the equation:

$$q''_s = q''_g + q''_{cs} + q''_{fr} - q''_l.$$

Before the pyrolysis front feeding heat flow is expended for the change of surface layer enthalpy of the material, $q''_h$. Radiation energy, absorbed by FHSM, was estimated by the difference between $q''_h$ and ($q''_g + q''_{cs}$). As it is seen from Figure 5, the main contribution into the change of surface layer enthalpy near the flame edge is made by heat transfer by radiation from flame and through condensed phase. Far from the edge this contribution is mainly made by the radiation from flame. The contribution of heat transfer through the gas phase near the flame edge is 9.5% from the total heat flow. The contribution of radiating energy is 49.1%, and that of heat transfer through condensed phase is 41.4%.

Apparently, sufficient contribution into total heat transfer of radiation heat transfer is connected with the sufficient carbon-back formation during FHSM combustion.

The character of radiative heat flow change in preflame zone in the distance of 5 mm from flame edge testifies that in this field the contribution of heat losses is visible, resulting reflection and energy irradiation from material surface.

The existence in PESC materials of metal oxides with high coefficients of radiating energy reflection in IR wave range [7] leads to the increase of reflective ability of FHSM.

The result obtained testify the complex mechanism of heat shield action of FHSM, based on PESC. Here the important role is played by both physical and chemical factors. Processes, proceeding in the

                    *M.A. Shashkina et al.*

condensed and gas phase at the interface, influences the character of heat and mass transfer at FHSM combustion and its fire shield functions.

The performed investigation of features of PESC compositions thermal transformations, the mechanism of fire and heat shield action of material, the influence of components on exploitational properties allowed to create rubber-like impact-resistant highly efficient FHSM.

Covers for FIRS, tested under the conditions, modeling natural ones, were processed from the material, worked out on the basis of PESC.

The following features were estimated during the test: stability for the influence of distributed static load of 1000 kG during 30 min along two axes of device; that of load of 250 kG, falling from 1 m at touch square of 1.6 cm$^2$ and blow overloads up to 1000 units.

It was found that there were no breaches of FHSM completeness observed during all these tests. The film with information is fully protected, and record can be played without disturbances.

The influence of thermal impact of 1000°C during 15 min, treating 100% of the surface, was estimated on FIRS construction of two types: i) usual construction with internal heat isolation, and ii) construction without it. Testing results of the first FIRS construction is the cover of the material, based on PESC and traditionally used rubber-tissue FHSM, are shown in Table 5. The following parameters were controlled during testing: Temperature at titanium FIRS box, time of reaching maximal temperature. The temperature after 15 min was measured at titanium box and inside FIRS container. Maximal temperature inside the container was measured also.

Table 4 shows the advantages of the material worked out, based on PESC. Tests of devices of the second construction possess the temperature of 150-180°C inside the container after 15 min influence in dependence on PESC composition content. However, temperature was being increased after the end of external heat influence, and was found 190-280°C after 20-22 min. Thus, high level of heat shield properties of the materials is found in this case, also.

**Table 4.**  Testing results of thermal impact stability for FIR with different fire and heat shield coatings.

| Shield Material | Temperature, °C | | | | |
|---|---|---|---|---|---|
| | on the outside titanium FIR box | | | inside FIR box | |
| Silicon Rubber Textile | 890 | 7$^X$ | 810$^X$ | 80$^X$ | 420 |
| Material based on PESC | 380 | 34 | 248 | 25 | 250 |

X - high temperature action has been stopped after achievement of temperature inside FIR box of 80°C.

The results obtained show, that worked out efficient FHSM, based on PESC gives the possibility to abandon complex FIRS constructions, meaning the application of internal thermal insulation.

## REFERENCES


1. Pat. USA 3934066.
2. J.A. Mansfield, S.R. Riccitiello, L.L Fewell, *J. Fire and Flammability* vol. 6, N4, 492 (1975).
3. Pat. Germany 1270792.
4. Pat. USA 3888557.
5. *Fire* vol. 69, N854, 135 (1976).
6. Pat. USA 3816226.
7. M. Derybers, *Practical Application of Infrared Beams*, Energy, Moscow (1959) (in Russian).
8. Yu.V. Polezhaev, F.B. Yurevich, *Heat Shield*, Energy, Moscow, (1976) (in Russian).
9. I.V. Margolin, N.P. Rumyantsev, *Principles of Infra Red Techniques*, Military Publishers (1957) (in Russian).
10. J. Lecont, *Infra red Radiation*, Phys. Matem. Publishers, Moscow (1958) (in Russian).
11. H.S. Carslaw, J.C. Jaeger, *Conduction of Heat in Solids*, Oxford University Press, London (1959).
12. D.E. Cagliostro, S.R. Riccitiello, K.J. Clark, A.B. Shimuzu, *J. Fire and Flammability*, vol. 205, 221 (1975).
13. J. Buckmaster, C. Anderson, A. Nachman. *Internat. J. Eng. Sci.*, vol. 24, 263 (1986).
14. A.M. Kanury, D.J. Holve, *J. Heat Transfer*, vol. 104, 338 (1982).
15. A.M. Bulgakov, V.I. Kodolov, A.M. Lypanov, *Combustion Modelation of Polymer Materials*, Chemistry, Moscow (1990) (in Russian).
16. R.M. Aseeva, G.E. Zaikov, *Combustion of Polymer Materials*, Hanser, Munich (1986).
17. T. Kashiwagi, 1987 Fall Technical Meeting Eastern States, Section of the Combustion Institute, Nov. 26 (1987), NBS, Geithersberg, M.D.
18. N.A. Khalturinsky, Al. Berlin, *Degradation and Stabilization of Polymers* by Ed. H.H.G. Jellinek, Elsevier, Amsterdam, vol. 2, ch. 3, 145 (1989).
19. R.M. Aseeva, G.E. Zaikov. *Developments in Polymer Stabilization* by Ed. G. Scott, Elsevier Appl. Sci. Publ. London, vol. 7, Ch. 5, 233, (1984).
20. L.A. Lovachev, *Academy Sciences News*, USSR, Ser. Chem., NI, 220 (1980).

21, **C.J. Quinn, G.H. Beall**, Fourth Annual BCC Conference on Flame Retardancy: Recent Advances in Flame Retardancy of Polymeric Materials, Stamford, Connecticut, May 18-20 (1993).

22. **J. Green**, *J. Fire Sci.*, vol. 10, Nov/Dec, 470 (1992); Thermoplastic Polymer Additives/Theory and Practice, by Ed. J.T. Lutz, Marcel Deccer, NY, ch. 4, 93 (1989).

23. **P. Hartey**, Des. Eng., oct., 81 (1972).

24. Pat. USA 3878167.

25. Pat. USA 3874889.

26. Japan Pat. 26246, cl. 25 (1) A, 261 (CO8R).

27. Pat. USA 3819577.

28. **B.N. Dolgov**. *Catalysis in Organical Chemistry*, Chemistry Publ., Moscow (1949) (in Russian).

29. **G. Camino, L. Costa, M.P. Luda**, Fourth Annual BCC Conference on Flame Retardancy: Recent Advances in Flame Retardancy of Polymeric Materials, Stamford, Connecticut, May 18-20 (1993).

30. **A.A. Donstov, G.A. Lozovik, S.P. Novytskaya**. *Chlorinated Polymers*, Chemistry, Moscow (1979) (in Russian).

31. **G.M. Ronkin**, *Plastmassy* (N8, 16 (1980).

32. **A.S. Levin**, *Chlorinated Polyethylene and its Application*, by Ed. A.M. Savransky, Scientific Research Institute of Technical-Economic Investigations for Textile Industry, Moscow (1973).

33. **V.I. Abramov, A.A. Krasheninnikova**, *Industry of chlorinated Polymers abroad*, Sci. Research Inst. Techn. -Econ. Invest. for Chemistry, Moscow (1978).

34. Pat. USA 2640048.

35. Pat. USSR 583139 (1976).

36. Pat. USA 3347835.

37. Pat. USSR 364634 (1970).

38. Pat. USSR 547456 (1974).

39. Pat. Canada 471037.

40. *Chemie et Industrie* vol. 51, N5, 452 (1969).

41. apan Pat. 27605 (1968).

42. *Industrial Chloro-organical Products*, by Ed. L.A. Oshin, Chemistry, Moscow (1978).

43. *Chem. Rundschau*, vol. 20, N9, 145, 147, 149 (1967).

44. *Rubber Age*, vol. 100, N10, 47 (1968).

45. *Express Inform. Synthetic High Molecular Materials* N2, 13 (1976) (in Russian).

46. Polyethylene sulphochlorinated Technical Condition N6-01-715-80, Ministry of Chem. Industry (1980) (in Russian).

47. **G.M. Ronkin, M.A. Korotyansky, A.I. Gershenovich, R.V. Dzhagatspanyan**, *Caoutchouc and Rubber* N1, 5 (1980) (in Russian).

48. **G.M. Ronkin, M.A. Korotyansky, E.S. Balakryrev, R.V. Dzhagatspanyan**. *Plastmassy*, N3, 26 (1980).

49. **O.V. Startsev, Yu.M. Vapyrov, A.S. Ovanesov, A.A. Donskoy, M.A. Shashkina**. *High molecular Compounds*, vol. 29A, N12, 2473 (1987).
50. *Caoutchouc and rubber* N1, 5 (1980) (in Russian).
51. **G.M. Ronkin**, *Chlorosulphonated Polyethylene*, Sci. Inv. inst. Techn. Econom. Inf. for Oil Chemistry Industry, Moscow (1977).
52. **V.G. Dyunina**. *Investigation in area of Production porous sole Material based on chlorosulphonated Polyethylene*, Dissertation Textile Institute, Moscow (1972).
53. **V.G. Dyunina, S.A. Pavlov, B.A. Dinzburg**. Leather-Shoes Industry, N5, 115 (1970) (in Russian).
54. **G.A. Blokh**. *Organic Accelerators for vulcanization of caoutchoucs*, Chemistry, Moscow (1964) (in Russian).
55. **G.A. Blokh**, *Organic Accelerators of vulcanization and vulcanization systems for Elastomers*, Chemistry, Leningrad (1978).
56. **I.I. Tugov, G.I. Kostyrina**, *Chemistry and Physics of Polymers*, Chemistry, Moscow (1989), p. 431, (in Russian).
57. **A.A. Dontsov, S.P. Novytskaya, I.M. Kochanov, L.N. Stepanova**. *Caoutchouc and Rubber* N3, 30 (1976).
58. **V.S. Kuzin, A.A. Dontsov, G.M. Ronkin, M.A. Korotyansky**. ibid N11, 14 (1975).
59. **A.A. Dontsov**. *High. Molecular Compounds*, vol. 18A, N1, 169 (1976).
60. **A.A. Dontsov, S.P. Novytskaya, L.N. Stepanova**. *Caoutchouc and Rubber*, N4, 24 (1977).
61. **A.F. Nospycov, G.A. Blokh**, ibid, N12, 18 (1983).
62. Pat. USSR 992539, KJI, CO8L 29/34.
63. **Z.A. Kovacheva, G.A. Zhuravleva, R.G. Komyadyna, N.D. Trufanova**. *Caoutchouc and Rubber* N11, 19 (1983).
64. **L.T. Goncharova, A.G. Shvarts, V.A. Sapronov**. ibid, N9, 18 (1983).
65. Elastomer-Berichte-72, DuPont Elastomer Hypalon. LD-974-10 Verzogerer Aktivator fur Hypalon; Chemische Bestendigkeit von Hypalon Vernetzungsprodukten.
66. **A.L. Labutin**. *Caoutchoucs in anticorrosive technics*, Chemistry Publ., Moscow (1962).
67. Pat. USA 3883615.
68. Japan Pat. 57-129682, CO8L 23/08; declar 59-20341 (1982).
69. Japan decl. 59-18744 (1982), CO8L 23/06.
70. Japan decl. 40-14803 (1965).
71. FRG decl. 3042089 (1981).
72. France decl. 2419957 (1979(, CO8L 23/28.
73. Japan decl. 54-26258 (1979).
74. **A.A. Donskoy, M.A. Shashkina, R.M. Aseeva, L.V. Ruban**. *Intern. J. Polym. Materials* vol. 24, N1-4, 157 (1994).

75. **R.M. Aseeva, L.V. Ruban, S.Kh. Korotkevich, A.A. Molchanov, G.E. Zaikov,** Plastmassy N8, 78 (1989) (in Russian).
76. **V.V. Gobdanova, I.A. Klymovtsova, B.O. Phylonov, S.S. Phedeyev** a.o.. *High molecular compounds* vol. 28, B, N1, 42 (1987) (in Russian).
77. **S.S. Phedevyev, V.V. Svyrydov, V.V. Bogdanova, A.Ph. Surteyev** a.o. reports of Byelarus. *Academy of Sciences,* vol. 27, N1, 56 (1983).
78. **L. Costa, G. Camino et al.** *Polymer Degrad. and Stability* vol. 30, 13 (1990).
79. **V.V. Bogdanova, I.A. Klymovtsova, S.S. Fedeev, A.I. Lesnikovich.** *Intern. J. Polymer Materials* vol. 2, 51 (1993).
80. **J.W. Hastie.** *Combustion and Flame,* vol. 77A, N6, 733 (1973).

# New Pathway of the Polymers Flammability Depression with the Polyvinyl Alcohol Oxidative System

**S.M. Lomakin, M.I. Artsis and G.E. Zaikov**
*Institute of Chemical Physics, Russian Academy of Sciences*
*4, Kosygin Street, Moscow 117334, Russia*

**Abstract** — Composition of Nylon 6,6 & Poly(vinyl alcohol) oxidized with the Potassium Permanganate in water solution

## 1 GENERAL DATA

### 1.1.  Polyvinyl Alcohol [1]

The monomer, vinyl alcohol ($CH_2$=CHOH, tautomeric with acetaldehyde, $CH_3CHO$), is unknown, but the polymer is a stable substance obtained by hydrolysis or alcoholysis of polyvinyl acetate,

$$-CH_2-CH- \quad \xrightarrow{\ H_2O \text{ or } ROH\ } \quad -CH_2-CH- \ + \ CH_3COOH$$
$$\quad\quad | \quad\quad\quad\quad\quad\quad\quad\quad\quad\quad\quad\quad\quad | \quad\quad (\text{or } CH_3COOR)$$
$$\quad OCOCH_3 \quad\quad\quad\quad\quad\quad\quad\quad\quad\quad OH$$

Soluble in water, especially on warming to 70-80°C. Plasticized by moisture and water-soluble hydric alcohols (glycerol, glycol, triethylene glycol, sorbitol), glyceryl lactate, amides (urea), and also calcium chloride acting as a humectant. Swollen by polyhydric alcohols (hot). Relatively unaffected by almost all organic liquids: hydrocarbons, and chlorinated hydrocarbons, alcohol, acetone, esters, oils and fats. Ozone - resistant. Decomposed by conc. acids, especially oxidizing acids. Combustion - burns with slightly smoky flame, leaves black residue and unpleasant odor (particularly formaldehyde).

Specific gravity 1.25-1.35, fibers 1.26-1.32., Thermal properties - Second-order transition at 85°C. The polymer is serviceable up to 120°-140°C through it slowly yellows (degrades) above 100°C, and darkens (with evolution of water and conversion to an unsaturated polymer) if kept at 150-200°C. The polymer does not properly melt.

The first-order transition temperature (melting transition) is considerable and depends on the tacticity of the PVA Table 1. [2].

**Table 1.** Crystalline melting points of PVA

| Experimental method | $T_m$ (°C) | Sample |
|---|---|---|
| DTA | 228-240 | Atactic |
| DTA | 230-267 | Syndiotactic |
| DTA | 212-235 | Isotactic |

Insolubilized fibers show changes associated with decomposition and melting between 200-260°C.

## 1.2 NYLON 6,6 [1]

$$\text{HOOC}(CH_2)_4\text{COOH} + H_2N(CH_2)_3NH_2 \xrightarrow[220° - 270°]{-H_2O} [-NH(CH_2)_6NH-CO(CH_2)_4CO-]$$

## PROPERTIES

Solvents - Phenols and phenol/chlorinated hydrocarbon mixtures (1/3 phenol/tetrachloroethane, by volume), 90% formic acid, hot formamide, hot benzyl alcohol, halohenated alcohols. Relatively unaffected by hydrocarbons and chlorinated hydrocarbons, esters, ethers, oils and set. Decomposed by conc. mineral acids, oxidizing agents, halogens. Combustion - melts, darkens, boils, finally burns with a small flame that is easily extinguished giving a white smoke. Decomposes 260-300°C.

Specific gravity 1.14. Thermal properties - second-order (glass) transition at 37-47°C. Melting point 250-260°C.

# 2 DECOMPOSITION

## 2.1 Thermal Decomposition of PVA [2]

### 2.1.a Pyrolysis under vacuum

Decomposition of PVA proceeds in two stages. The first stage, which begins at 200°C, mainly involves dehydration accompanied by the formation of some volatile products. The residues are predominantly polymers with conjugated unsaturated structures. In the second stage, polyene residues are further degraded at 450°C to yield carbon and hydrocarbons. The mechanism involved in thermal decomposition of PVA has been deduced by Tsuchya and Sumi [6]. At 245°C water is split off the polymer chain, and a residue with conjugated polyene structure results:

$$(-CH-CH_2)_n-CH-CH_2- \longrightarrow (-CH=CH_2)_n-CH-CH_2- + nH_2O$$

Scission of several carbon-carbon bonds leads to the formation of carbonyl ends. For example, aldehyde ends arise from the reaction:

$$-CH-CH_2-(CH=CH)_n-CH-CH_2- \longrightarrow -CH-CH_2-(CH=CH)_n-CH + CH_3-CH-$$

In the second-stage pyrolysis of PVA, the volatile products consist mainly of hydrocarbons, i.e. n-alkanes, n-alkenes and aromatic hydrocarbons (Table 2) [6].

**Table 2.** Thermal decomposition products of PVA (240°C, 4h)

| Product | % by weight of original polymer |
|---|---|
| Water | 33.4 |
| CO | 0.12 |
| $CO_2$ | 0.18 |
| Hydrocarbons $C_1$ - $C_2$ | 0.01 |
| Acetaldehyde | 0.17 |
| Acetone | 0.38 |
| Ethanol | 0.29 |
| Benzene | 0.06 |

**Table 1** (Continued)

| | |
|---|---|
| Crotonaldehyde | 0.76 |
| 3-pentene-2-one | 0.19 |
| 3,5-heptadiene-2-one | 0.099 |
| 2,4-hexadiene-1-al | 0.55 |
| Benzaldehyde | 0.022 |
| Acetophenone | 0.021 |
| 2,4,6-octatriene-1-al | 0.11 |
| 3,5,7-nonatriene-2-one | 0.020 |
| Unidentified | 0.082 |

It was reported also [2] that the formation of volatile vapors, water and acetaldehyde from PVA pyrolyzed at 185-350°C can be fitted into the first-order rate equation. The rate constant is equal to $10^{-4}$ per minute.

## 2.1.b Pyrolysis in the presence of oxygen

Thermal degradation of PVA in the presence of oxygen can be adequately described by two-stage decomposition scheme, with one modification. Oxidation of unsaturated polymeric residue from dehydration reaction introduces ketone groups into polymer chain. These groups then promote the dehydration of neighboring vinyl alcohol units producing conjugated unsaturated ketone structure [2]. The first-stage degradation products of PVA pyrolyzed in air are fairly similar to those obtained in vacuum pyrolysis. In the range of 260-280°C, the second-order-reaction expression satisfactorily accounts for the degradation of 80% hydrolyzed PVA up to a total weight loss of 40%. The activation energy of decomposition appears to be consistent with the value of 53.6 kkal/mol which is obtained from the thermal degradation of PVA [2].

## 2.1.c Infrared spectroscopic study of the PVA residue

It was studied the changes in the IR spectra of PVA subjected to heat treatment [2]. After heating at 180°C in air the bands appeared at 1630 cm$^{-1}$ (c=c stretching in isolated double bands), 1650 cm$^{-1}$ (c=c stretching in conjugated diens and triens), and 1590 cm$^{-1}$ (c=c stretching in polyenes). The intensity of the carbonyl stretching frequency at 1750-1720 cm$^{-1}$ increased, although the rate of increase of intensity was less than of the polyene band at low temperatures. Above 180°C, although dehydration was the predominant reaction at first, the rate of oxidation increased after an initial induction period.

## 2.1.d Mechanism of crosslinking during degradation of PVA

The identification of a low concentration of benzene among the volatile products of PVA [81, 84] has been taken to indicate the onset of a crosslinking reaction proceeding by a Diels-Alder addition mechanism [86]. Clearly benzenoid structures are ultimately formed in the solid residue, and the IR spectrum of the residue also indicated the development of aromatic structures [85].

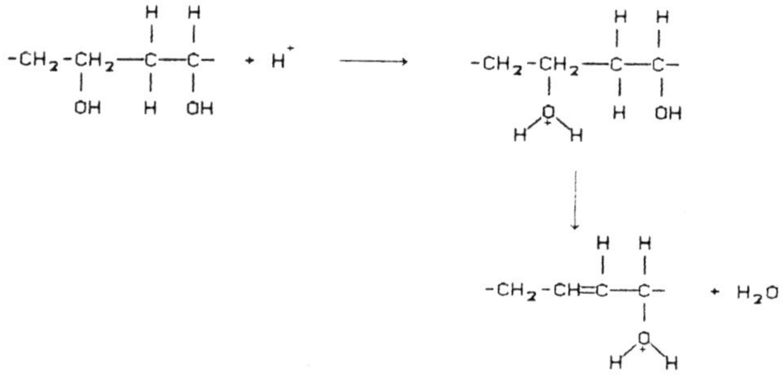

Acid-catalyzed dehydration promotes the formation of conjugated sequences of double bonds (a) and Diels-Alder addition of conjugated and isolated double bonds in different chains may result in intermolecular crosslinking producing structures which form graphite on carbonization (b)

(a)

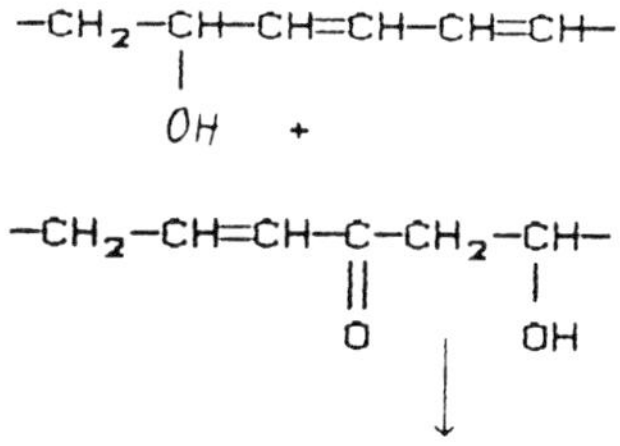

(b)

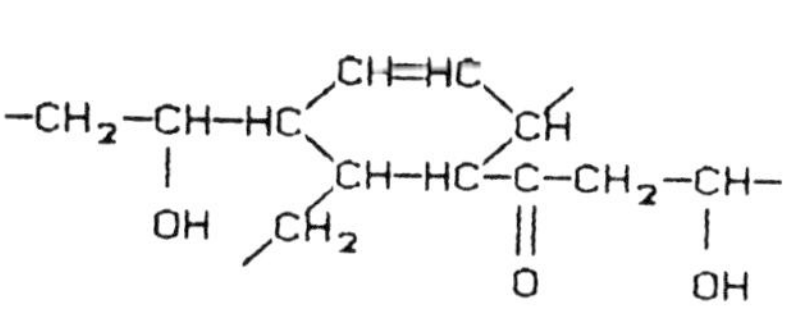

## 2.2 Thermal Decomposition of NYLON 6,6

It was found [3,4] that the NYLON 66 was subjected to temperatures above 300°C in an inert atmosphere it completely decomposed. The wide range of degradation products, which included several simple hydrocarbons, cyclopentanone, water, CO, $CO_2$ and $NH_3$ suggested that the degradation mechanism must have been highly complex. Further research has led to a generally accepted degradation mechanism for the aliphatic polyamides [5]:

$$-[-\overset{O}{\overset{\|}{C}}-(CH_2)_x-\overset{O}{\overset{\|}{C}}-NC-(CH_2)_y-NH-]_n \xrightarrow{H_2O} -[\overset{O}{\overset{\|}{C}}-(CH_2)_x-\overset{O}{\overset{\|}{C}}-OH + H_2N-(CH_2)_y-NH-]$$

$$\Big\downarrow \Delta \qquad\qquad \Big\downarrow \Delta \qquad\qquad \Big\downarrow \Delta$$

$$-[-\overset{O}{\overset{\|}{C}}-(CH_2)_x \quad \overset{O}{\overset{\||}{C}} \quad NC-(CH_2)_y \quad NH]- \qquad -[\overset{O}{\overset{\|}{C}}-(CH_2)_x \quad CO_2$$

$$NH_3 \quad (CH_2)_y \quad NH]$$

Hydrocarbons
Cyclic ketones, esters
Nitriles, Carbon char

1. Hydrolysis of the amide bond occurred below the decomposition temperature and could have been alleviated if proper care had been taken in purification;

2. Homolytic cleavage of C-C, C-N, C-H bonds generally began at the decomposition temperature and occurred simultaneously with hydrolysis;

3. Cyclization and homolitic cleavage of products from both of the above reactions occurred;

4.   Secondary reactions produced CO, $NH_3$, nitriles, hydrocarbons, and carbon chars.

# 3. Fire Retardancy of Polyamide 6,6

The fire retardancy of polymers can be achieved by different ways:

1. Modifying the pyrolysis scheme: in order to produce non volatile, or non combustible products which dilute the flame oxygen supply.

2. Smothering the combustion through dilution of the combustible gases, or the occurrence of the barrier (char) which hinders the supply of oxygen.

3. Trapping the active radicals in vapor phase (and eventually in condensed phase).

4. Reducing the thermal conductibility of the material in order to limit the heat transfer (char).

In Polyamide 6,6 (NYLON) several types of flame retardants are used [7]:

Phosphorus compounds

Halogenated compounds with inorganic synergists

Miscellaneous compound (non phosphorus and non halogenated)


**Phosphorus compounds** retard burning essentially in the condensed phase, influencing pyrolysis and char formation, according to the following mechanism [8]:

In the condensed phase:

1. They form phosphoric and related acids which act as "heat sink" as they undergo endotermic reduction.

2. They form a thin glassy coating which is a barrier that lowers the concentration of combustible gases in vapor phase; limits the diffusion of oxygen; limits the heat transfer.
In vapor phase, where PO radicals are likely to exist:

3. They stop the free-radical oxidation process of carbon into carbon monoxide only, avoiding the highly exothermic reaction of carbon dioxide formation.

**Halogenated compounds** are known to be essentially active in vapor phase by trapping the active radicals, and so inhibiting the combustion of the matrix.

There is a well-known synergy between $Sb_2O_3$ and the halogenated compounds: $Sb_2O_3$ promotes the halogen volatilization in the form of metallic halides $(SbX_3)$ or oxyhalides $(Sb_xO_yX_z)$. In the case of chlorinated compounds $SbCl_3$ traps H to produce HCl, SbCl, $SbCl_2$ and Sb. Sb reacts with O and OH to give SbOH and SbO which traps radicals again [9].

**Miscellaneous systems** concerns the flame retardants which act through the production of diluting non-combustible gases:

1. Hydroxides decompose endothermically, thus the heat level, producing water which dilutes the combustible gases, and inert oxide. In the case of $Mg(OH)_2$, it has been reported that its morphology has an influence on the LOI but no effect on the UL94 test [10]. It is also claimed that $Mg(OH)_2$ promotes char formation and gives lower smoke levels than $Al(OH)_3$ [11].

2. Melamine cyanurate retards burning in the vapor phase by producing uncombustible diluting gases $(NH_3)$ and condensation products in the condensed phase, which undergo further degradation steps [12].

## 3.1 Industrial Flame Retardants for NYLON 6,6

1. Red phosphorous (6-9% wt.) in reinforced and non-reinforced formulations (0-35% wt. glass fibers) [7].

2. Dechlorane (10-25% wt.) $C_{10}Cl_{12}$ [7].

3. Decabromodiphenyl oxide (10% wt.) $C_{12}OBr_{10}$ [7].

4. Decabromodiphenyl $C_{12}Br_{10}$ or brominated polystyrene (7-24% wt.), in synergy with antimony oxide (0.6 to 6% wt.) or a mixture of antimony oxide with zinc borate in various proportions, in reinforced and non-reinforced formulations (0-35% wt.) [7].

5. Halowax 1006 (a mixture of penta and hexachloronaphtalene, with 63% chlorine) - 10% wt. with: SnO 10% wt. [13],

    PbO 10% wt. [13],

    ZnO 15% wt. [13].

6. Archlor 1268 (chlorinated biphenyl - 68% chlorine) - 10% with $SnO_2$ - 10% wt. [13].

7. Pentabromophenol - 5% wt. with PbO - 5% wt. [13].

8. Octachloronaphthalene - 5% wt. with $Cu_2O$ - 3% wt. [13].

9. Melamine cyanurate (10% wt.) [7].

10. Magnesium hydroxide (50-60% wt.) is used in reinforced formulations [7].

## 3.2. REQUIREMENTS FOR FLAME RETARDED FORMULATIONS BASED ON NYLON 6,6

| Property | Requirement |
| --- | --- |
| Thermostability | > Nylon processing T° (300°C) |
| Volatility | Non volatile below 300°C |
| Flame retardancy | LOI≥28, UL94: V0 or V1 |
| Mechanical properties | Impact strength ≥ 40 kJ/m$^2$ |
| Hazard | Safe handling, no toxicity, no pollution |
| Possibility of filler | Yes |
| Cost | Not expensive |

# 4. Oxidation of Poly (Vinyl Alcohol) by Permanganate Ion in Alkaline Solutions

The literature available on the oxidation of macromolecules by alkaline permanganate presents little information about these redox systems. It was investigated [14, 15] the oxidation of PVA as a polymer containing secondary alcoholic groups by $KMnO_4$ in alkaline solution. It was reported that the oxidation of PVA by $MnO_4$ in alkaline solutions occurs through formation of two intermediate complexes (1) and/or (2) [15]. Hence, two alternative mechanisms for the decomposition were suggested. The first corresponds to a fast deprotonation for intermediate by the alkali followed by electron transfer from PVA$^-$ to $Mn^6$ (a). The second mechanism corresponds to the transfer of hydride ion from PVA to $MnO_4^{2-}$ as the rate determining steps (b):

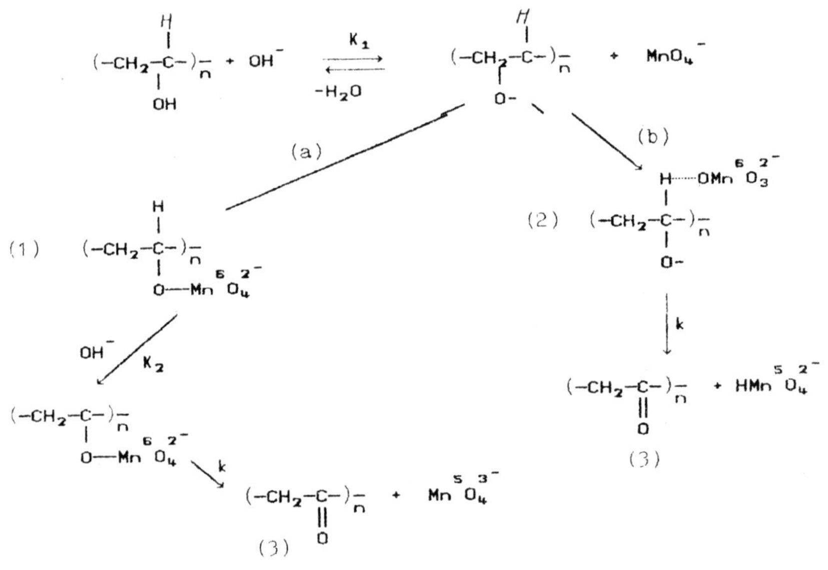

The reactions (a) and (b) lead to the formation of poly(vinyl ketone) (3) as a final product of oxidation of the substrate. Poly (vinyl ketone) was isolated and identified by microanalysis and spectral data [16].

## REFERENCES


1. *Handbook of Common Polymers*, by J.R. Scott, W.J. Roff, J. Pacitty, CRC Press, 1973, 688 p.
2. *Polyvinyl Alcohol. Properties and Applications*, ed. by C.A. Finch, John Wiley & Sons, London-New York-Sydney-Toronto, 1973, 622 p.
3. **B.G. Achhammer, F.W. Reinhard, G.M. Kline**, *J.Appl. Chem.*, v.1, 1951, p. 301.
4. **I. Goodman**, *J. Polymer Sci.*, v. 13, 1954, p. 175.
5. *Thermal Stability of Polymers*, ed. by R.T. Conley, Marcel Deccer Inc., New York, 1970, v. 1, p. 350.
6. **Y. Tsuchiya, K. Sumi**, *J. Polym. Sci.*, A-1, v. 7, p. 3151, 1969.
7. **S. Bodrero**, Makromol. *Chem., Macromol. Symp.*, v. 74, 1993, p. 137.
8. **M. Lewin, S.M. Atlas, E.M. Pearce**, *Flame Retardant Polymeric Materials*, Plenum Press, 1975.

9.  **J.W. Lyons**, *The Chemistry and Uses of Fire Retardants*, Robert E. Krieger Publishing Company, 1987.
10. **D. Sallet, V. Mailhos-Lefierre, B. Martel**, *Polym. Degrad. and Stab.*, v. 30, 1990, p. 73.
11. **L. Keating, S. Petric, G. Beekman**, *Plast. Compd.*, 9(4), 40, 1986.
12. **L. Costa, G. Camino**, *Journ. of Thermal Anal.*, v. 34, 1988, p. 423.
13. *Flame Retardant Polymers*, M.W. Ranney, Noyes Data Corporation, Park Ridge, New Jersey, USA, 1970, 254 p.
14. **R.M. Hassan**, *Polymer International*, v. 30, 1993, pp. 5-9.
15. **R.M. Hassan, S.A. El-Gaiar, A.M. El-Summan**, *Polymer International*, v. 32, 1993, pp. 39-42.
16. **R.M. Hassan, M.A. Abd-Alla, M.A. El-Gahmi**, *J. Mater. Chem.*, v. 2, 1992, p. 613.

# Pyrolysis and Carbonization of Cross-Linked Poly (Methyl Metacrylate)


*Sergei M. Lomakin and Guennady E. Zaikov*
*(Institute of Chemical Physics, Moscow 117977, Russia)*



**Abstract—** The thermal degradation of network copolymers of methyl methacrylate was studied as a function of the chemical nature of the cross-linking agent and the frequency of cross-links. Unlike the linear homopolymer, both the trimethylolpropane triacrylate and trimethylolpropane trimethacrylate networks were found to char when burned. The differences in the thermal degradation of these polymers are interpreted in terms of a single model for kinetics of depolymerization.


## INTRODUCTION

Our research has focused on finding ways to increase the tendency of plastics to char when they are burned. There is a strong correlation between char yield and fire resistance [1]. This follows because char is formed at the expense of combustible gases and because the presence of a char inhibits further flame spread by acting as a thermal barrier around the unburned material. The tendency of a polymer to char can be increased with chemical additives and by altering its molecular structure. We have identified some factors which promote the formation of char. Here, we extend this research to consider the thermal degradation behavior and flammability of chemically cross-linked poly (methyl methacrylate) (PMMA).

## EXPERIMENTAL

PMMA samples were synthesized by thermal radical polymerization of MMA in the presence of BPO (1.5% wt.). MMA was dissolved in toluene to produce a 50% by weight solution. This solution was heated at 60°C for 3 hours. When this solution started to become viscous it was poured into hexane. Then, after origin a solid phase of PMMA it was dried and solid polymer was put into Petri dishes and cured for 72 hours at 80°C and after that for 6 hours at 120° to remove additional solvent [2]. The number of average molecular weight ($M_n$) and polydispersity obtained from size exclusion chromatography (in THF) were 2.9-3.9 x $10^4$, respectively. The two cross-linked PMMA networks were prepared in toluene solution by free radical copolymerization of MMA with trimethylolpropane triacrylate (TTA) and trimethylolpropane trimethacrylate (TTM). The actual cross-link densities of the copolymers were determined by solvent swelling in toluene and acetone using the technique described in references [3,4],

$$-\ln(1 - v_r) - v_r - \chi_1 v_r^2 = \rho V_o \left( \frac{1}{M_c} - \frac{2}{M} \right) \left( v_r^{1/3} - v_r/2 \right)$$

where $v_r$ - volume fraction of polymer in the swollen system,

$\chi_1$ - polymer/solvent interaction parameter,

$\rho$ - density of unswollen polymer,

$V_o$ - molar volume of solvent,

$M_c$ - average mol. wt. of polymer between cross-links,

M - average mol. wt. before cross-linking.

The solvents in which the network was swollen to equilibrium were toluene with $V_o$=106.3 cm$^3$/mol and $\chi_1$=0.45 (25°, 27°C) and acetone with $V_o$=77.3 cm$^3$/mol and $\chi_1$=0.48 (17°C) [3].

The cross-link density (a) was defined as $1/2M_c$. For an appropriate result, if M is much greater than $M_c$ and if swelling is considerable, the negative terms on the right-hand side of the equation can be neglected, since then 2/M is small compared with $1/M_c$, and $v_r/2$ is small compared with $v_r^{1/3}$ [4].

The Final Swollen Equilibrium Polymer Volume Fraction, $v_r$, was determined as

$$v_r = \frac{W_o \rho_s}{(W_1 - W_o) \rho_p} \times 100\%$$

where $W_o$ - weight of polymer before swelling,

$W_1$ - weight of polymer after swelling,

$\rho_p$ - density of polymer,

$\rho_s$ - density of solvent.

Thermogravimetric analysis (TGA) were performed in a nitrogen atmosphere using a Mettler TA 2000 thermoanalyzer at a slow heating rate (2°C/min). Char yield measurements were made under controlled conditions using the NIST Cone Calorimeter [5]. The samples (10-20 g) were placed in dishes which were exposed to a radiant heat flux of 20 $kW/m^2$.

## RESULTS AND DISCUSSION

It is well known that depolymerization to the monomer is a major reaction path in the low temperature thermal degradation of PMMA [6]. The derivative thermogram of linear PMMA (Figure 1) indicates at least three reaction channels. The peaks centered at 200, 270 and 350°C have been previously assigned to head-to-head (H-H), and chain (E) and random (S) scission initiation respectively [6].

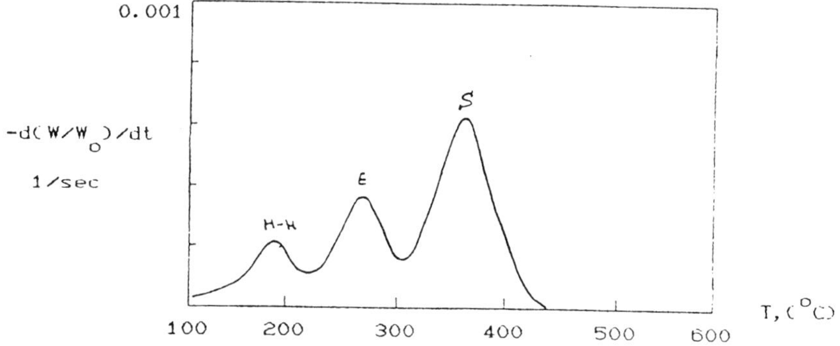


**Figure 1.** The DTG of linear PMMA.

It is well known a simple kinetic model for global depolymerization can be based on the assumption that the total number of radical polymer fragments (R) attains a steady state. The rate of weight loss from the condensed phase polymer ($W_o$-W) should be proportional to the rate at which monomer molecules $\lambda(P_1)$ are formed [7].

$$dM_1/dt=dP_1/dt=-(k_id_o/m_o+k_t\beta)R_1-k_p(R-R_1) \tag{1}$$

$P_n$ and $R_n$ - the number of polymer molecules and free radical polymer fragments having degree of polymerization n, $m_o$ - molecular weight of the monomer, V is the volume occupied by the condensed phase polymer, $M_1=\Sigma nP_n=W/m_o=d_oV/m_o$, the rate constants for intermolecular radical transfer: $P_n+R_s \rightarrow P_r+R_{n-r}+P_s$; termination: $R_n \rightarrow P_n$ ($\beta=1$) or $R_r+R_{n-r} \rightarrow P_n$ ($\beta=R/V$); and depropagation: $R_n \rightarrow R_{n-1}+P_1$ are $k_1$, $\beta k_t$ and $k_p$, respectively.

The first term on the right hand side of Equation 1 corresponds to the rate of formation of monomer from radical transfer and termination. The second term represents the rate of formation of monomer from depropagation. Radical transfer and termination account for a relatively small fraction, $\approx x^{-1}$ (where x is the average degree of polymerization), of the total weight loss. Consequently, the contribution from the first term to the integrated area of the peak in the derivative thermogram of a depolymerizing sample will be negligible compared to the contribution from the second term and can, therefore, be ignored.

The net rates of changes in the number of radicals having degree of polymerization n=1 and n≥2 are

$$dR_1/dt-(2k_s+k_iRd_o/M_1m_o+k_e) \sum_{j=2}P_j+k_2P_2-(k_id_o/m_o+k_t\beta)R_1+k_pR_2 \qquad (2)$$

$$dR_n/dt=(2k_s+k_iRd_o/M_1m_o) \sum_{j=n+1}P_j+k_eP_{n+1}-(k_id_o/m_o+k_p+k_t\beta)R_1+k_pR_{n+1} \qquad (3)$$

where $k_s$ denotes the rate constant for scission of c-c bonds at random positions along the polymer chain and $k_e$ is the rate constant for the scission of a vinyl end-group. Setting the sum over n equal to zero gives the following equation for the number of radicals at the steady state:

$$0=(2k_s+k_iRd_o/M_1m_o) \Sigma(n-1)P_n+2k_e\Sigma P_n-(k_iRd_o/m_o+k_t\beta)R \qquad (4)$$

The solution for x>>1 is

$$R = \frac{2(k_s + k_e/x)M_1}{\beta k_t} = [2(k_s + k_e/x)/k_t(m_o/d_o)^{N-1}]^{1/N} M_1, \qquad (5)$$

where N=1 for first order termination and N=2 for second order termination.

After substituting into Equation (1) for R from Equation (5), we obtain

$$dM/dt=-2(k+k/x)\gamma^{-1}M \qquad (6)$$

where the average number of monomer molecules which unzip before termination $(\gamma^{-1})$ is less than the number average degree of polymerization (x).

The peaks corresponding to the individual reactions which contribute to the total rate of depolymerization can be resolved by dynamic heating to the extent that $sk_e(T)k_s(T)d_t$ approaches zero. Assuming this to be the case, the global activation energies for Random and End-Chain initiation corresponding to this model are

$$\Delta E_s= \begin{cases} E_s+E_p-E_t & \text{(first order termination)} \\[4pt] (E_s-E_t)/2+E_p & \text{(second order termination)} \end{cases} \tag{7}$$

and

$$\Delta E_e= \begin{cases} E_e+E_p-E_t & \text{(first order termination)} \\[4pt] (E_e-E_t)/2+E_p & \text{(second order termination)} \end{cases} \tag{8}$$

where $E_s$, $E_p$, $E_t$ and $E_e$ are the activation energies for the elementary reactions.

An expression for the average rate of weight loss can be derived from the following argument: at any given moment there is a total of $R/2$ activated polymer molecules. Each of these fragmented polymer molecules will unzip an average of x molecules during the time it takes to terminate depropagation, $(\beta k_t)^{-1}$. After substituting for R from Equation (5) we obtain,

$$dM_1/d_t=R_x\beta k_t/2=-(k_s+k_e/x)xM_1 \tag{9}$$

This expression for the rate of weight loss indicates that the derivative thermogram of PMMA should have two peaks.

The high temperature reaction is Random scission initiated depolymerization, which should have an activation energy of

$$\Delta E_s=E_s$$

At lower temperatures, only End-Chain initiation can occur.

$$\Delta E_e=E_e$$

The generalization of Equations (6) and (9) to account for other forms of initiation, such as the scission of Head-to-Head (h-h) bonds in PMMA, is straightforward. Thus, when $\gamma^1>x$, there will be additional peak in the derivative thermogram corresponding to the term $-2(k_{h\text{-}h}/x)\gamma^1M_1$ with

$$\Delta E_{h\text{-}h}= \quad E_{h\text{-}h}+E_p-E_t \text{ (first order termination)}$$

$$(E_{h\text{-}h}\text{-}E_t)/2+E_p \quad \text{(second order termination)}$$

and when $\gamma^1 < x$, there should be a peak corresponding to the term $-k_{h\text{-}h}M_1$ with $\Delta E_{h\text{-}h}=E_{h\text{-}h}$.

Cross-linked PMMA may be viewed as an inter-connected network of linear chains. We assume that the average kinetic chain length ($x_1$) of cross-linked polymer molecules is comparable to the average degree of polymerization in the linear polymer. If the cross-linking agent does not interfere with depropagation, then the rate of depolymerization should be given by

$$dM_1/d_t=Rx\beta k_t/2=-(k_s+k_e/x)xM_1 \quad \text{with } x=x_1$$

On the other hand, if the cross-linking agent obstructs further depropagation then x should be replaced by $N_c$, the average number of monomer molecules per cross-link.

This analysis suggests that it may be possible to stabilize PMMA and other polymers which degrade by depolymerization as well, by copolymerization.

The features present in the derivative thermograms of TTM-co-PMMA network copolymers (Figure 2) are qualitatively similar to those observed for linear PMMA.

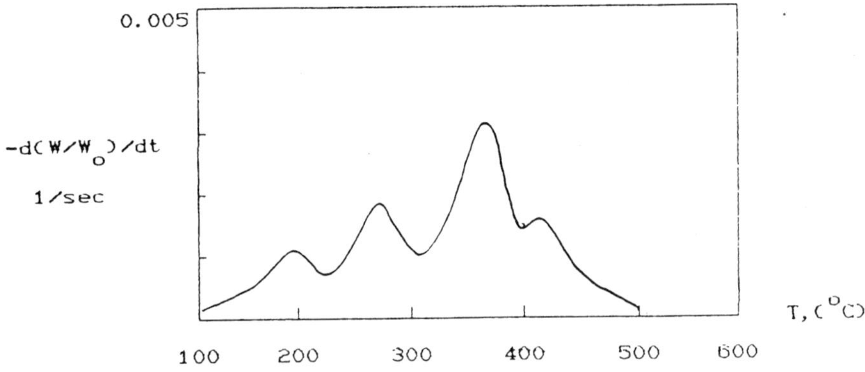


**Figure 2.** The DTG of PMMA-co-TTM 56:1

In contrast to this behavior, there are difference between the derivative thermograms of TTA-co-PMMA network copolymers (Figure 3) and PMMA. The peaks corresponding to H-H and end-chain initiated depolymerization are less pronounced in the derivative thermograms of the TTA-co-PMMA network copolymers.

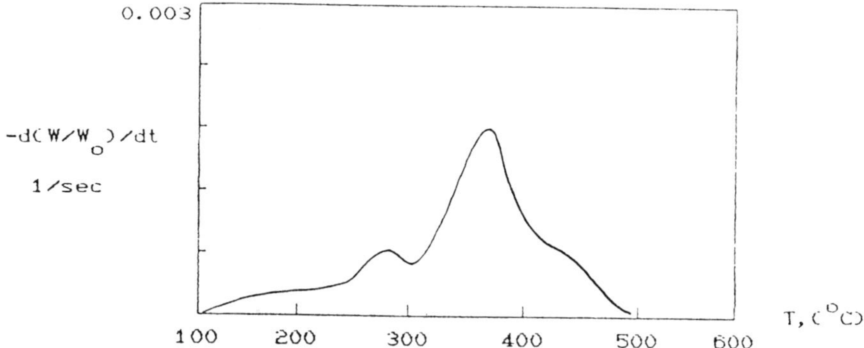


**Figure 3.** The DTG of PMMA-co-TTA 45:1

Our explanation for this behavior is based on the recognition that acrylate polymers, unlike the corresponding methacrylates, do not depolymerize [8]. The assumption is that the presence of a TTA in the network suppresses further depropagation so that the average zip length of TTA/MMA copolymer is equal to the average number of monomer units between cross-links ($N_c$). As a result, each initiation is less productive in TTA/MMA than it was in either TTm/MMA or PMMA and the H-H and end-chain sites are depleted without significant weight loss. The loss in intensity of the peaks corresponding to the low temperature depolymerization of the TTA/MMA network copolymers is accompanied by an increase in the peak attributed to random scission initiated depolymerization. This is a necessary consequence of the fact that the polymer chains which survive H-H and end-chain initiated depolymerization are susceptible to depolymerization initiated by random scission.

The activation energy for random scission initiated depolymerization should increase when cross-links are present because breaking a bond in the interior necessarily places an additional strain on the other bonds in the network (Table 1,2). The Arrhenius constants for copolymer thermal degradation were calculated using the Kissinger method [9].

$$\ln(b/T^2_m=)\ln(AR/E)-E/RT_m$$

where: b - the heating rate, $T_m$ - temperature of the maximum of the DTG peak, E - effective activation energy, A - pre-exponential factor

**Table 1.**   Effective kinetic parameters of the thermal degradation of cross-linked PMMA-co-TTM

| PMMA-co-TTM 56:1 | | $A, s^{-1}$ | $E, kJ/mole$ |
|---|---|---|---|
| 1 | peak | $0,77 \times 10^{11}$ | 152 |
| 2 | peak | $1,73 \times 10^{15}$ | 218 |
| 3 | peak | $1,30 \times 10^{16}$ | 263 |
| 4 | peak | $0,75 \times 10^{15}$ | 276 |

**Table 2.**   Effective kinetic parameters of the thermal degradation of cross-linked PMMA-co-TTA

| PMMA-co-TTA 45:1 | $A, s^{-1}$ | $E, kJ/mole$ |
|---|---|---|
| 1st peak | – | – |
| 2nd peak | $0,50 \times 10^{14}$ | 203 |
| 3rd peak | $3,50 \times 10^{16}$ | 270 |

Measured values of the global activation energies are surprisingly low for scission of the corresponding C-C bonds ($\approx 335$ kJ/mole). The source of this discrepancy may be that mass transport of the degradation products plays an important role in determining the rate of weight loss. This effect would be expected to play an increasingly important role in determining the kinetics of weight loss from more cross-linked polymers. On the other hand, it is also possible that the activation energies are low because the radical character of the

**Table 3.**   Char yield after TGA analysis (600°C)

| Polymer | | % wt. char residue |
|---|---|---|
| PMMA | | – |
| PMMA-co-TTM | 26:1 | – |
| PMMA-co-TTM | 56:1 | – |
| PMMA-co-TTM | 82:1 | – |
| PMMA-co-TTM | 125:1 | – |
| PMMA-co-TTA | 20:1 | 1.4 |
| PMMA-co-TTA | 45:1 | 0.8 |
| PMMA-co-TTA | 78:1 | – |

degrading polymers is never fully developed in the transition state. The radical sites formed by scission of σ bonds between monomer units are made by concurrent formation of π bonds as they propagate down the chain. The derivative thermograms of some of the more highly cross-linked polymers exhibit peak centered at about 420-450°C. This peak was present in the derivative thermograms of all the polymers which produced char when they were burned, while it is absent from the thermograms of those polymers which did not char. Furthermore there is a positive correlation between the intensity of this peak and the amount of char produced when the polymer was burned (Table 3,4).

**Table 4.** Char yield after Cone Calorimeter tests

| Polymer | | % wt. char residue |
|---|---|---|
| PMMA | | – |
| PMMA-co-TTM | 50:1 | 0.32 |
| PMMA-co-TTA | 26:1 | 1.42 |
| PMMA-co-TTA | 32:1 | 0.32 |
| PMMA-co-TTA | 62:1 | – |

The copolymerization of MMA with TTM does not appear to offer a significant improvement in thermal stability. The more highly cross-linked PMMA-co-TTM, however, do produce some char when they are burned. The fact that TTA itself does not depolymerize has a dramatic effect on the thermal degradation behavior of the PMMA-co-TTA network copolymers. The presence of a TTA node in the network suppresses further depolymerization so that each initiation is less productive than it was in either PMMA-co-TTM or PMMA. Indeed, the low temperature reaction channels (H-H and end-chain initiated depolymerization), which play a prominent role in the thermal degradation of linear PMMA, are almost completely shutdown in the more highly cross-linked PMMA-co-TTA. We have already alluded to the fact that the rate of weight loss due to random scission initiation depolymerization is dramatically reduced in the most highly cross-linked TTA copolymers. These observations indicate that it may be possible to improve thermal stability by substituting MMA/acrylate copolymers for PMMA in some applications. The char forming tendency of the cross-linked copolymers should contribute to a further reduction in the flammability of these materials.

# REFERENCES


1. **Van Krevelen**, *Polymer*, 16, 615 (1975).
2. **Singh, S., Frisch, H.L., Ghiradella, H.** *Macromolecules* 23, 375 (1990).
3. *Encyclopedia of Polymer Science and Engineering*, 13, John Wiley and Sons, New York, 453 (1976).
4. **Miller, J.H., Leung, H.,** *J. Macromol. Sci. Chem.*, A4, 1705 (1990).
5. Standard Test Method for Heat and Visible Smoke Release Rates for Materials and Products Using an Oxygen Consumption Calorimeter, ASTM: Philadelphia, PA, 1-17 (1991).
6. **Grassie, N. Melville, H.W.,** Proc. Roy. Soc. (London), A199, 14 (1949).
7. **Boyd, R.H.,** in *Thermal Stability of Polymers*, Conley, R.T., Ed. Marcel Dekker, Inc., New York, 47 (1970).
8. **Lomakin, S.M., Aseeva, R.M., Zaikov, G.E.,** *Pol. Degr. and Stab.*, 36, 187 (1992).
9. **Kissinger, H.E.,** *Anal. Chem.*, 29, 1702 (1957).

# Polypropylene Flame Retardant System

## Si-SnCl$_2$ - Flame Retardant Composition


*G.E. Zaikov, S.M. Lomakin and M.I. Artsis*
*Institute of Biochemical Physics*
*Russian Academy of Sciences, Moscow 117334*


The subject of ecological safeness of polymer flame retardants has become a major problem in modern polymer industry. The different types of polymer flame retardants based on halogens (Cl, Br), heavy and transition metals (Zn, V, Pb, Sb), phosphorus-organic compound may cause an elimination of hazard products during polymer combustion and pyrolysis.

The fire retardancy of polymers can be achieved by different ways:

1. Modifying the pyrolysis scheme: To produce non volatile, or non combustible products that dilute the flame oxygen supply.

2. Smothering the combustion through dilution of the combustible gases, or the occurrence of the barrier (char) which hinders the supply of oxygen.

3. Trapping the active radicals in vapor phase (and eventually in condensed phase).

4. Reducing the thermal conductibility of the material to limit the heat transfer (char).

     *G.E. Zaikov, S.M. Lomakin, M.I. Artsis*

In our research we have focused on the ways 2,3 and 4. It has been proposed.

## Silicon-inorganic Flame Retardant Compositions - Gaseous/solid Phase Inhibitors of Polymer Combustion

The proposed flame retardant composition was based on the assumption of trapping the active radicals in vapor phase and eventually in condensed phase. The system we've proposed was silicon-inorganic composition (SI). This is one of the most interesting modern flame retardant systems. The possibility of reaction of gaseous-phase inhibition is based on the formation of SiCl, and HCl that can be produced only at temperatures above 300°-500°C [1]:

350°-500°C:     1.  $2 SnCl_2 + n\ Si = 2\ Sn + (n-1)Si + SiCl_4$
                     $SiCl_4 + 2\ H_2O = 4HCl + SiO_2$

410°C:          2.  $2\ PbCl_2 + n\ Si = 2\ Pb + (n-1)\ Si + SiCl_4$

280°-350°C:     3.  $4\ CuCl + nSi = 4Cu + (n-1)\ Si + SiCl_4$

300°C:          4.  $2\ CaCl_2 + nSi = 2\ Ca + (n-1)\ Si + SiCl_4$

400°C:          5.  $4FeCl_3 + nSi = 4\ Sn + (n-1)Si + 3SiCl_4$

In this temperature region $SiCl_4$ can depress the combustion of polymers in gaseous phase and also in the solid one. In solid phase $SiCl_4$ may react as "carbonizator" to produce a char. I gaseous phase $SiCl_4$ and HCl are the inhibitors of radical chain reaction of propagation in flame zone.

It has been directly chosen a SI-system with $SnCl_2$ for polyolephynes (polypropylene) because of the wide temperature range of reaction 350°-500°C. Presumably the same temperature range may be realized in solid phase of polymer during combustion.

# EXPERIMENTAL

## Materials

The polymer used in this work was Polypropylene, isotactic were supplied by Scientific Polymer Products, Inc., USA. The inorganic additives were, Stannous Chloride A.C.S. (Fisher Sci. Comp.), Sodium Borate, A.C.S. ($Na_2B_4O_7*10H_2O$, Fisher Sci. Comp.), Silicon, metal, 325 mesh, 99% (Aldrich Co.).

## Preparation of samples, incorporation of additives

Inorganic additives were mixed with Polypropylene powder in laboratory blender. The samples were prepared by press moulding at temperatures 120-140°C during 4 minutes.

## Cone Data PP-Silica Systems
### Heat flux=20.0 kW/m$^2$

| Pol System heat res., % | Initial wt., g kW/m$^2$ | Carbon rel., MJ/m$^2$ | Ignition t., s | Peak RHR, | Total |
|---|---|---|---|---|---|
| PP: Si/$Na_2B_4O_7$ 85: (10/5)  (a) | 42.0 | 12.4 | 131 | 428.9 | 341.33 |
| PP:Si/$Na_2B_4O_7$ /$SnCl_2$ 85:(8/5/2)  (b) | 41.5 | 28.7 | 394 | 432.3 | 194.36 |
| Polypropylene-iso Sp$^2$ | 21.7 | 0.3 | 217 | 1266.7 | 200.84 |

Cone Calorimeter tests of the polymer samples, as discs (radius 35 mm), were carried out at 20, 30, 35 and 50 kW/m$^2$. Each specimen was wrapped in aluminum foil and only the upper face was exposed to the radiant heater.

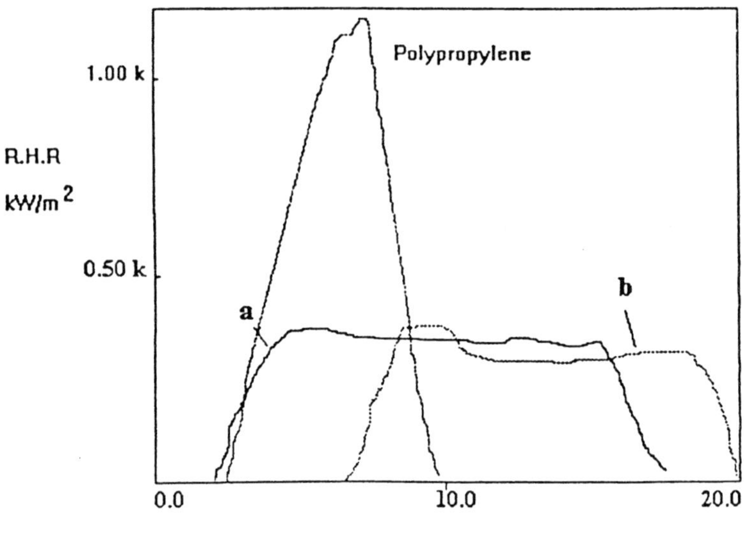


**Figure 1.** Rate of Heat Release vs. time for
Polypropylene;
Polypropylene with $Si/Na_2B_4O_7$ (85%:10%:5% wt.) - (a);
Polypropylene with $Si/Na_2B_4O$, $/SnCl_2$ (85%:8%:5%:2% wt.) - (b)
at heat flux 20kW/m².

# RESULTS AND DISCUSSION

The cone data (20 kW/m²) of polypropylene with Si-composition are given in Table 1 and Figure 1-6. The sodium borate (5 and 2% by wt.) was incorporated into these systems as intumescent agent. The addition of 10% wt. of silicon significantly depresses the flammability of polypropylene (Figure 1). However, the incorporation only 2% wt. of $SnCl_2$ has increased the ignition time delay in two times due to inhibition of combustion in gaseous phase by $SiCl_4$ and HCl.

The next set of Cone test was carried out at heat flux 35 kW/m² (Table 2, Figures 7-11). The flame retardant composition included 3% wt. of Si and 2% of $SnCl_2$. All Cone results indicate an improvement of fire resistance of Si-polypropylene composition in comparison with pure polypropylene Figures 7-11.

**Table 2.** Cone Data of Si-PP system at heat flux of 35 kW/m$^2$

| Cone Data | Polypropylene | PP+Si+SnCl$_2$ (95:3:2) |
|---|---|---|
| Char yield, % wt. | 0.0 | 10.1 |
| Ignition time, sec. | 62 | 91 |
| Peak RHR, kW/m$^2$ | 1378.0 | 860.1 |
| Total Heat Release, MJ/m$^2$ | 332.0 | 193.7 |

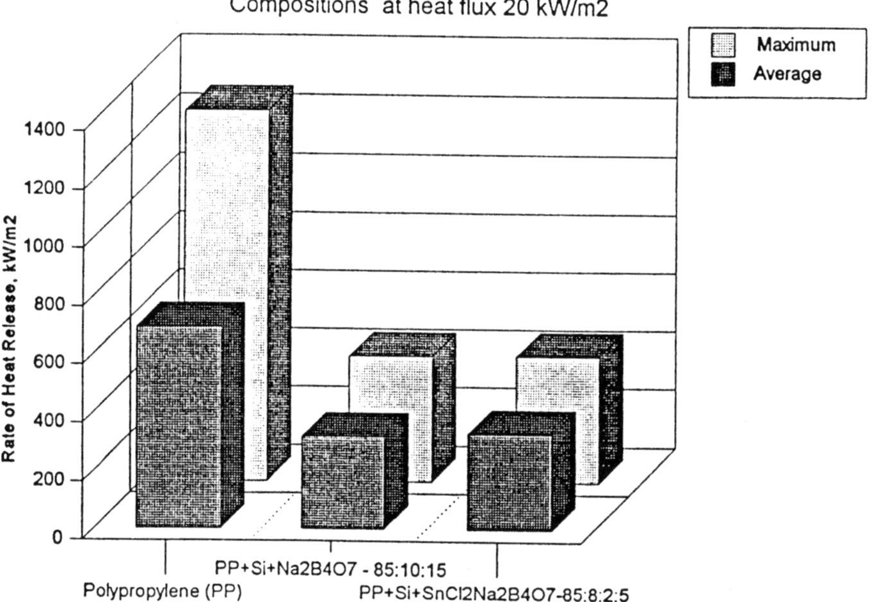


**Figure 2.** Cone Calorimeter Data of Polypropylene-Si compositions at heat flux of 20 kW/m$^2$ Rate of Heat Release

 *G.E. Zaikov, S.M. Lomakin, M.I. Artsis*

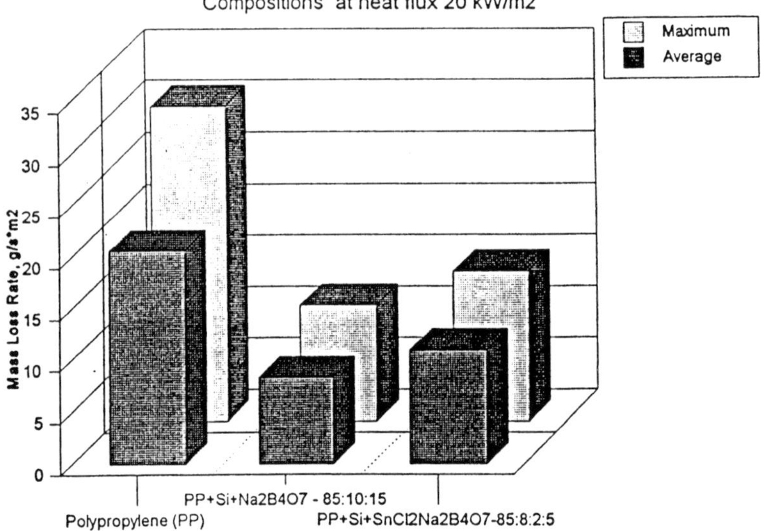


**Figure 3.** Cone Calorimeter Data of Polypropylene-Si compositions at heat flux of 20 kW/m$^2$ Mass Loss Rate.

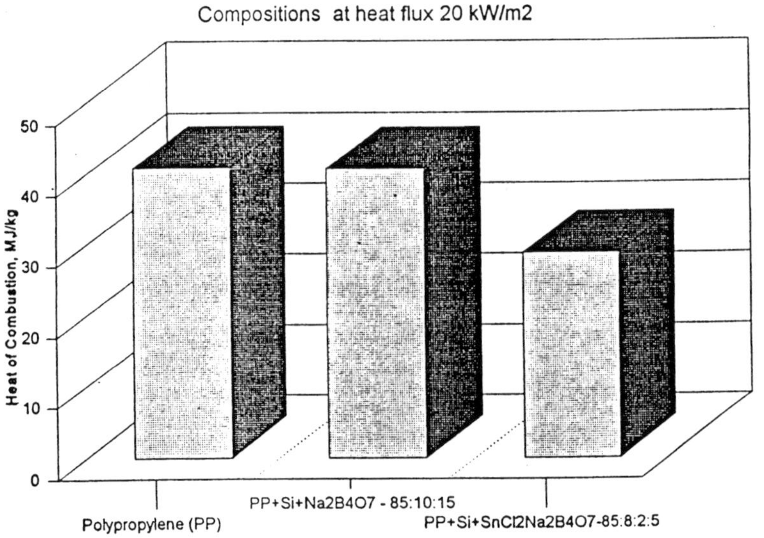


**Figure 4.** Cone Calorimeter Data of Polypropylene-Si compositions at heat flux of 20 kW/m$^2$ Heat of Combustion

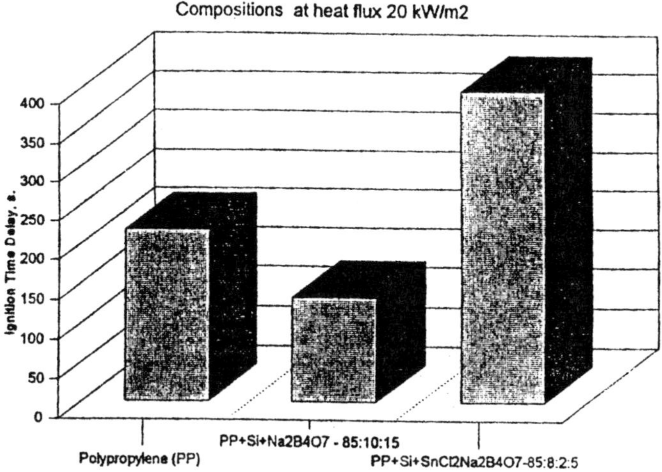



**Figure 5.** Cone Calorimeter Data of Polypropylene-Si compositions at heat flux of 20 kW/m² Ignition time

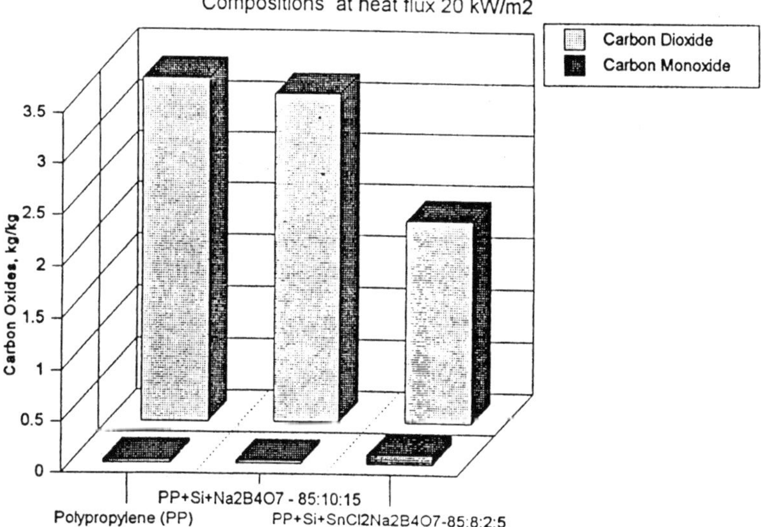



**Figure 6.** Cone Calorimeter Data of Polypropylene-Si compositions at heat flux of 20 kW/m² Carbon Oxides

 *G.E. Zaikov, S.M. Lomakin, M.I. Artsis*

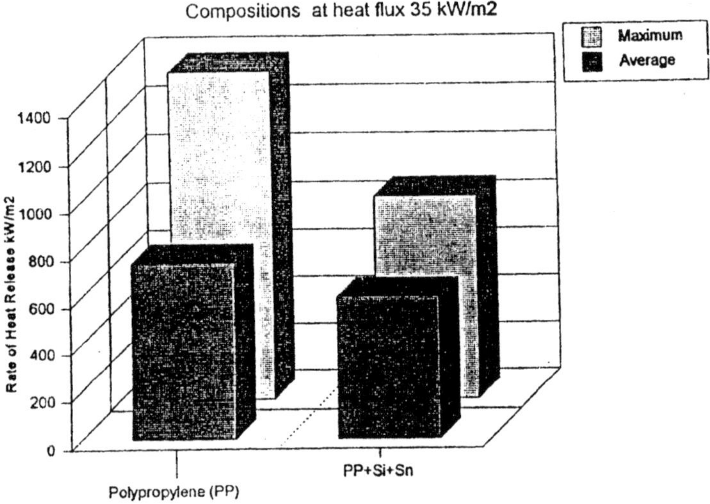



**Figure 7.**  Cone Calorimeter Data of Polypropylene-Si composition at heat flux of 35 kW/m$^2$ Rate of Heat Release

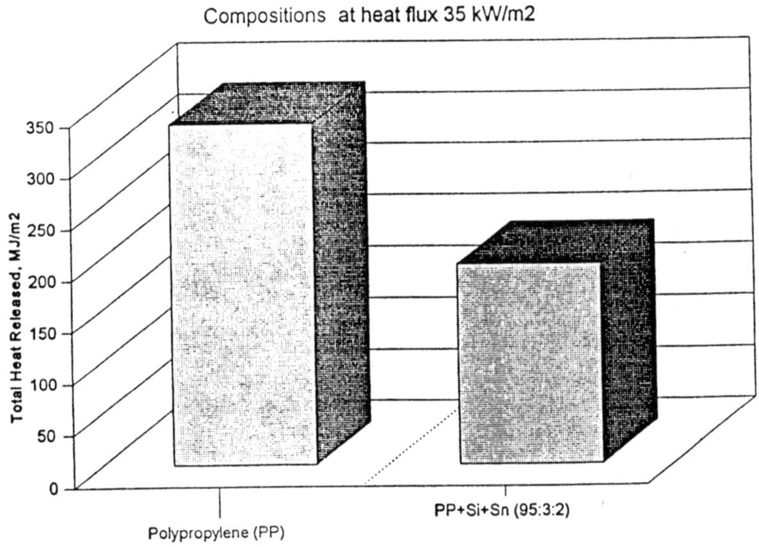



**Figure 8.**  Cone Calorimeter Data of Polypropylene-Si composition at heat flux of 35 kW/m$^2$ Total Heat Release

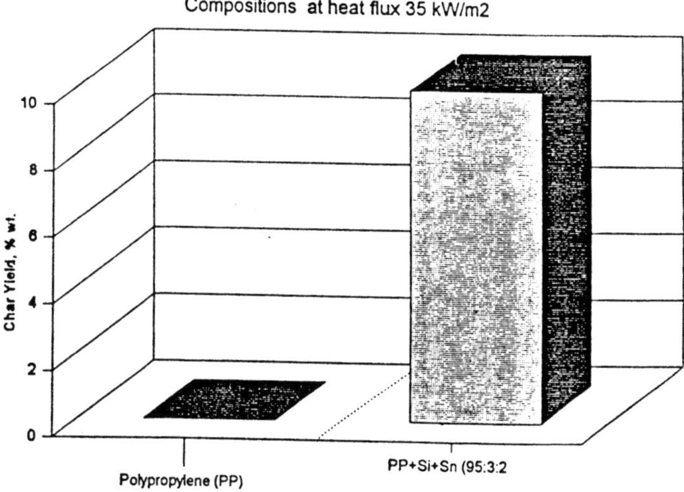


**Figure 9.** Cone Calorimeter Data of Polypropylene-Si composition at heat flux of 35 kW/m$^2$ Char Yield

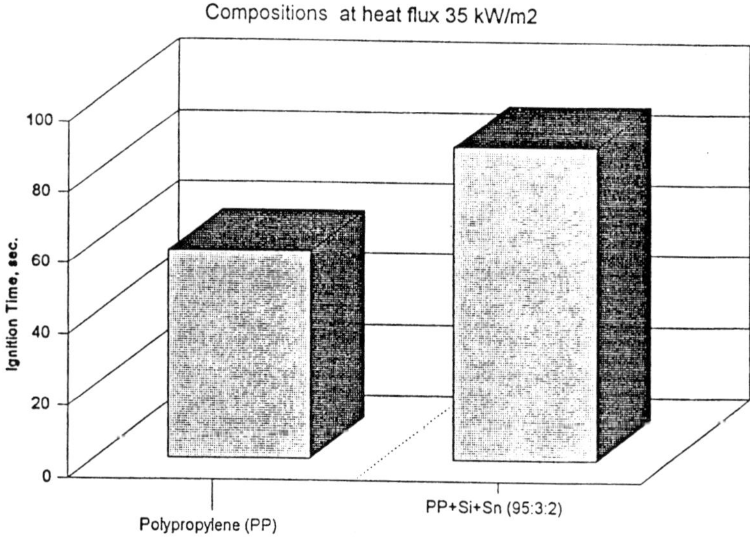


**Figure 10.** Cone Calorimeter Data of Polypropylene-Si composition at heat flux of 35 kW/m$^2$ Ignition time

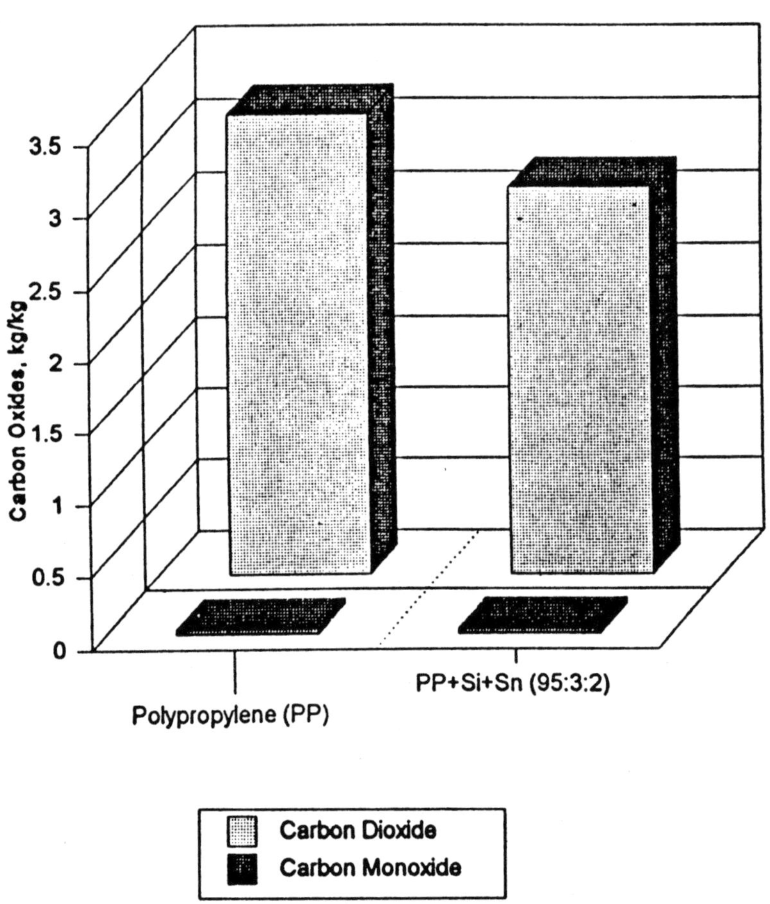



**Figure 11.**    Cone Calorimeter Data of Polypropylene-Si composition at heat flux of $35 \text{ kW/m}^2$ Smoke Area

## CONCLUSIONS

(1) The SI-system for polypropylene significantly suppressed the combustion of polypropylene. HCl and $SiCl_4$ inhibit the combustion in a gas phase and dramatically decrease the ignition time delay from 217 sec. to 394 sec. for polypropylene (20 kW/m$^2$, Table).

(2) Silicon-tin chloride polymer composition takes the "transition" place in the ecological scale of safety for flame retardants because of high temperature emanation of HCl and $SiCl_4$. However, these reactions take place only at temperatures above 300°C. While, at temperatures below 300°C SI-system is highly environmentally safe.

## REFERENCE

1. **R.J.H. Voorhoeve**, *Organohalosilanes, Precursors to Silicones*, Elsevier Co., 1967, 356 p.

# NEW TYPES OF ECOLOGICALLY SAFE FLAME RETARDANT SYSTEMS FOR PMMA


*G.E. Zaikov, S.M. Lomakin and M.I. Artsis*
*Institute of Biochemical Physics*
*Russian Academy of Sciences, 4 Kosgyin St., Moscow*


## INTRODUCTION

Thermal degradation of PMMA is the only source of fuel to sustain burning. It may also be the source of volatile decomposition products that quench flame, and it may lead to the formation of a char which impedes heat transfer back to polymer.

PMMA provides one of the simplest degradation routes. Following chain homolysis, the macroradical simply depropogates to monomer, a process sometimes described as "unzipping". The average zip length for the process is about 200 repeat units:

$$
\text{wwCH}_2-\underset{\underset{\text{COOCH}_3}{|}}{\overset{\overset{\text{CH}_3}{|}}{C}}-[\,\text{CH}_2-\underset{\underset{\text{COOCH}_3}{|}}{\overset{\overset{\text{CH}_3}{|}}{C}}-]_n-\text{CH}_2-\underset{\underset{\text{COOCH}_3}{|}}{\overset{\overset{\text{CH}_3}{|}}{C}}
$$

$$\downarrow$$

$$
\text{wwCH}_2-\underset{\underset{\text{COOCH}_3}{|}}{\overset{\overset{\text{CH}_3}{|}}{C}}-[\,\text{CH}_2-\underset{\underset{\text{COOCH}_3}{|}}{\overset{\overset{\text{CH}_3}{|}}{C}}-]_{n-1}\,\text{CH}_2-\underset{\underset{\text{COOCH}_3}{|}}{\overset{\overset{\text{CH}_3}{|}}{C}} \quad + \quad \text{CH}_2=\underset{\underset{\text{COOCH}_3}{|}}{\overset{\overset{\text{CH}_3}{|}}{C}}
$$

By this process, the polymer can provide an abundant supply of fuel for burning. PMMA doesn't produce any char (if it's pure).

Fire retardants incorporated into polymers can prevent or reduce the formation of fuel, they can quench flame, or they can make it more difficult for heat to be transferred back to the bulk of the polymer (e.g. causing a char form).

**Ammonium polyphosphate** is well known flame retardant for PMMA. On heating this loses ammonia, water to form first an ultraphosphate and then polyphosphoric acid, which at still higher temperatures fragments to give smaller phosphoric acid type fragments [1]. These species effect in the condensed phase, modifying the degradation route of the polymer. When ammonium polyphosphate is used as a fire retardant in PMMA, the effect is to greatly reduce the zip length and give in addition some char [1]. The action of Sandotlam 5085 fire retardant is similar, although the effectiveness is improved

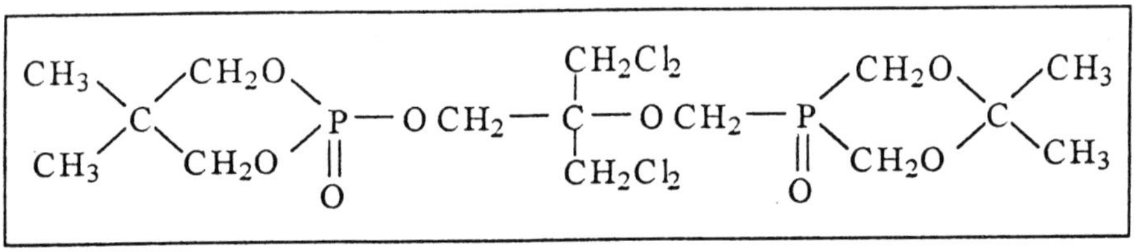

$$CH_3\diagdown \quad \diagup CH_2O\diagdown \qquad\qquad\qquad CH_2Cl_2 \qquad\qquad\qquad \diagup CH_2O\diagdown \quad \diagup CH_3$$
$$\phantom{CH_3}C\phantom{xxx} P-O\,CH_2-\overset{|}{\underset{|}{C}}-O\,CH_2-P\phantom{xxx}C$$
$$CH_3\diagup \quad \diagdown CH_2O \diagup \overset{\|}{\phantom{P}}\underset{O} \qquad CH_2Cl_2 \qquad \overset{\|}{\phantom{P}}\underset{O}\diagdown CH_2O \diagup \quad \diagdown CH_3$$

by the good compatibility. This compound breaks down thermally to give some traces of HCl and phosphoric acid type residue as the flame depressants. This acid may interact with the PMMA chain to convert some of esters' units to the cyclic anhydride structures [2].

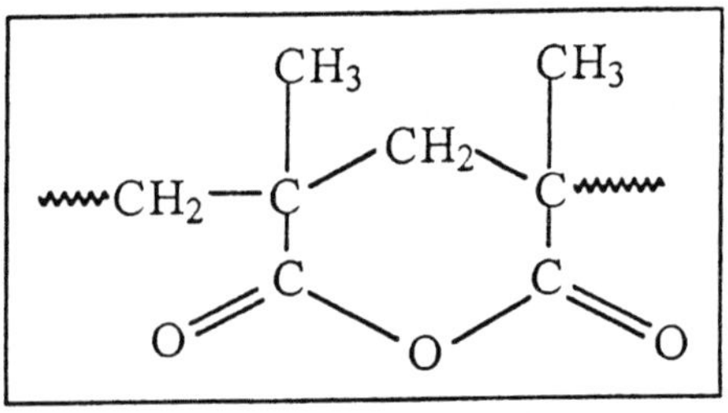

$$CH_3 \qquad\qquad CH_3$$
$$\sim\!\!\sim\!\!CH_2-\overset{|}{C}\diagup{}^{CH_2}\diagdown\overset{|}{C}\sim\!\!\sim\!\!\sim$$
$$O\!=\!\!C\diagdown\phantom{x}\diagup C\!\!=\!\!O$$
$$O$$

When the polymer chain begins to unzip following scission, the unzipping process stops at the first ring structure instead of continuing for 200 units. Further chain scission is needed more monomer, so the temperature of breakdown is raised significantly and furthermore, the anhydride rings result in some char residue. The halogen-containing degradation product may have a role in the gas phase in flame quenching. The other "types" of Sandoflams-Sandoflam 5086 and Sandoflam 5087 also have a halogens (Br and Cl) as the active components.

Another way of increasing the PMMA combustion characteristics is to alloy it with PVC [3].

Wilkie *at al.*, have demonstrated that certain transition metal compounds can induce char formation in PMMA [4-8]. There were founded some factors which are important in stabilization/destabilization of PMMA: (1) the acidity of metal ion; the strength of the bond between the transitionmetal and its counter-ion. An ion which is unable to coordinate to the polymer will not effect the degradation process. Ions which may coordinate and which also have relatively weak M-X bonds lead to destabilizing effects on the degradation, whereas ions which coordinate but have stronger M-X bonds have a stabilizing effect. However, this research still doesn't indicate a clear dependence of PMMA combustibility vs. transition metal compounds' incorporation.

The subject of ecological safeness of polymer flame retardants has become a major problem in the modern polymer industry. The different types of polymer flame retardants based on halogens (Cl, Br), heavy and transition metals (Zn, V, Pb, Sb) or phosphorus-organic compounds may reduce risk during polymer combustion and pyrolysis, yet may present ecological issues.

That is why, our study has been focused on the way to find out new types of ecologically safe flame retardant systems for PMMA. This goal might be achieving by means of the strong "char" barrier on the PMMA surface to heat and mass transfer process. This barrier needs to be coherent, fast-forming, but the barrier material need not be entirely carbonaceous.

Another approach to preventing the burning away of the char was to produce a ceramic layer containing (Si-C) bonds. In the work of *I. Janigova and I. Chodak* it was studied the crosslinking process of polyethylene by Silica Gel initiated by thermal decomposition of high temperature peroxide [9].

From all above the facts we proposed an ecologically safe flame retardant system for PMMA:

**Silica Gel-(3% wt.) + Luperox 2,5-2,5 (2,5-Dimethylhexane-2,5-dihydroperoxide)-(0.5% wt.)**

Half-life of "Luperox" at selected temperatures

| Temperature, °C | Half-life, hours |
| --- | --- |
| 145 | 19.0 |
| 160 | 6.1 |

## EXPERIMENTAL

**Materials.** The PMMA used in this work was supplied by Scientific Polymer Products, Inc., USA, Silica Gel - 28-200 mesh (Fisher Sci. Co.), High temperature peroxide - LUPEROX 2,5-2,5 (Luisidol Div., USA).

**Preparation of samples, incorporation of additive.** The samples for combustion measurements (PMMA, blend of PMMA, Silica Gel and Luperox 2,5-2,5 - 96.5%:3%:0.5% by wt.) were prepared in a laboratory blender at room temperature (10 min), the mixed samples were compression molded at temperature 80°C for 10 min. The weight of samples was $55^{\pm}0.3$ g.

**Cone Calorimeter tests** on the polymer samples, as discs (radius 35 mm), were carried out at 3 5 kW/m$^2$. Each specimen was wrapped in aluminum foil and only the upper face was exposed to the radiant heater.

## RESULTS AND DISCUSSION

Cone results (Figure 1-6) suggest an improvement of fire resistance of the PMMA-SG composition in comparison with pure PMMA. Char yield of PMMA-SG composition exceeds the amount of initial additives in twice (Figure 2). This is an evidence of carbonization process in this system during combustion. Rate of Heat release suggest improvement of fire resistance characteristics for PMMA-SG composition in comparison with PMMA (Figure 1,3), and Total Heat Release (Figure 4), Mass Loss Rate (Figure 5) and Heat of Combustion (Figure 6). However, a less satisfactory correlation is given in the determination of Ext. Smoke Area (Figure 7). Also the Cone measurements do not show an increase of Carbon monoxide (Figure 8).

## CONCLUSIONS

Proposed PMMA-SG composition represents one of the "transition" flame retardant system. Our final goal is to design an ecologically safe char/ceramic "former". It would be desirable to investigate more closely an influence of Silica Gel (in different compositions) on the polymer combustion.

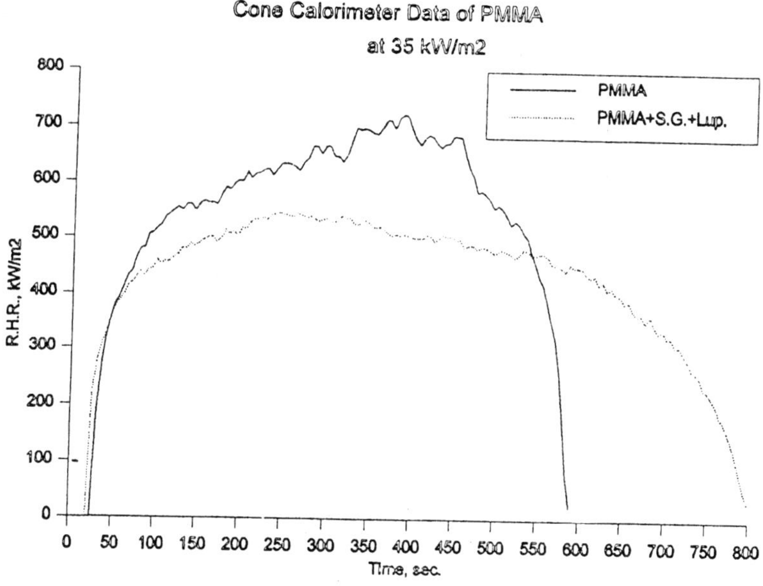



**Figure 1.**

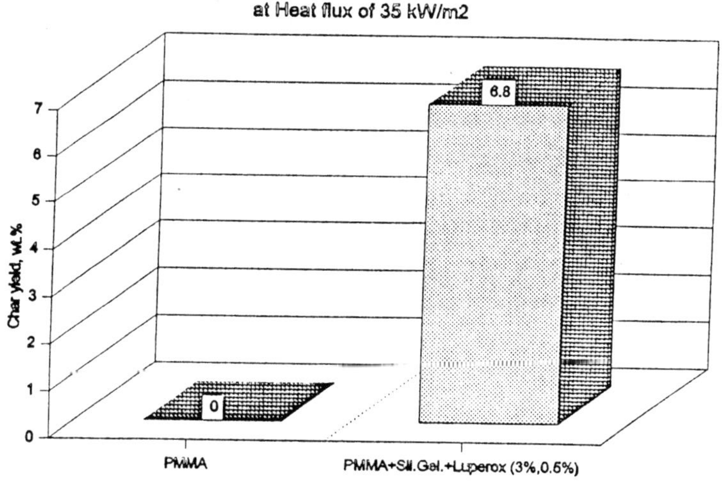



**Figure 2.**

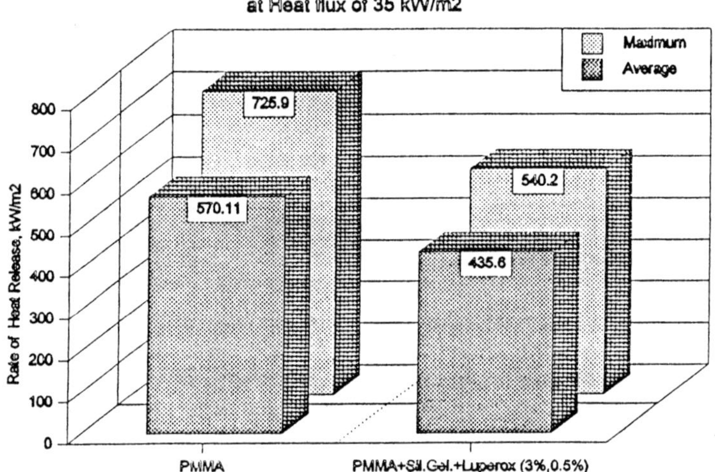



**Figure 3.**

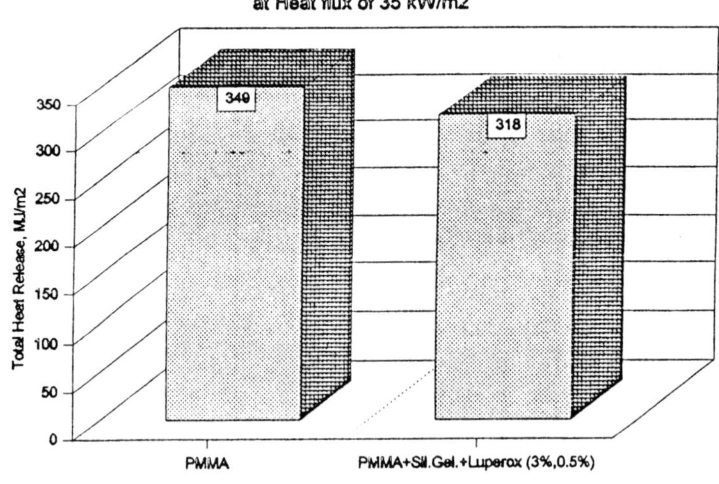



**Figure 4.**

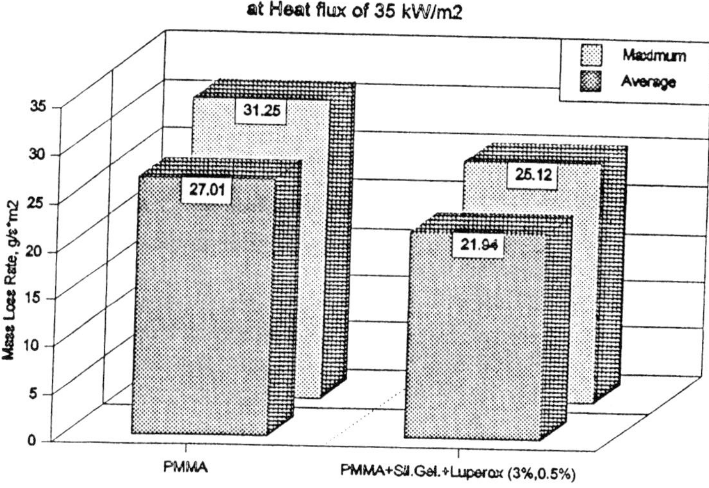


Figure 5.

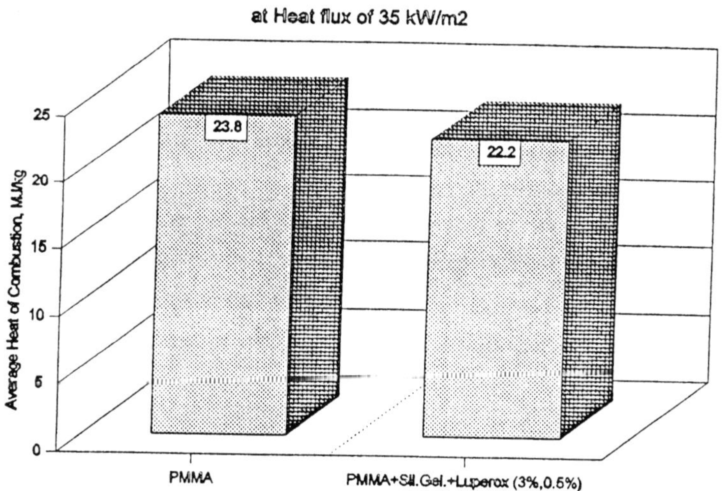


Figure 6.

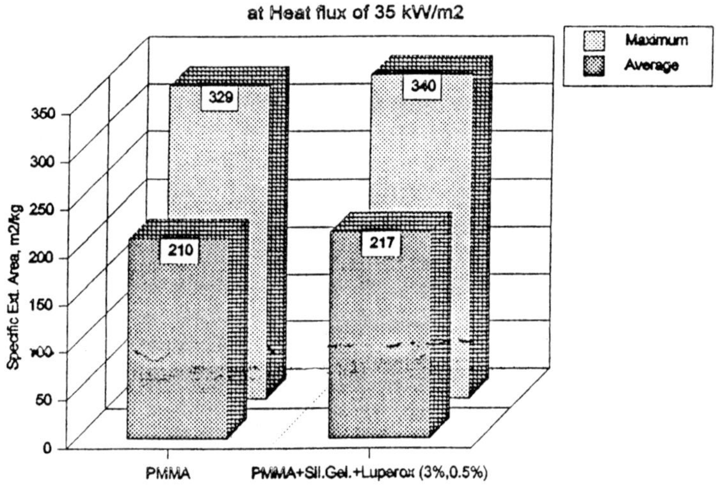



**Figure 7.**

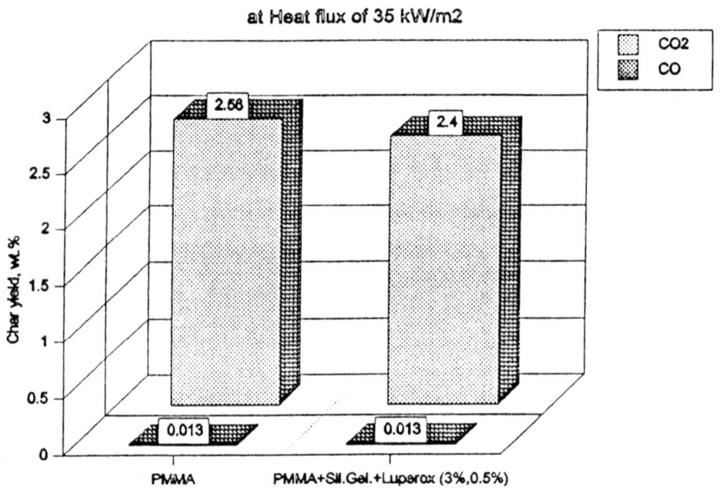



**Figure 8.**

# REFERENCES


1. G. Camino, N. Grassie, I.C. McNeill, *Journ. Pol. Sci.*, Pt. A1, 15, 1978, p. 2381
2. I.C. McNeill, *Macromol. Chem.*, Macromol. Symp., 74, 1993, p.11
3. Franc L. Fire, *Combustibility of Plastics*, VNR, New York, 1991, 319 p.
4. Wilkie, C.A., Sirdesai, S.J., Suebsaeng, T., Chang, P., *Fire Safety Journal*, 1991, 15, p. 297
5. J. A. Chandrasiri, Wilkie, C.A., *Journal Thermal Degradation and Stab.*, 1994, p. 83
6. J. A. Chandrasiri, Wilkie, C.A., *Journal Thermal Degradation and Stab.*, 1994, p. 91
7. J. A. Chandrasiri, D.E. Roberts, Wilkie, C.A., Journal Thermal Degradation and Stab., 1994, p.97
8. Wilkie, C.A., R. Scott Beer, J.T. Leone, *Journal of Fire Science*, 1993, 11, p. 184
9. I. Janigova, I. Chodak, *Eur. Pol. Journal*, 1994, v. 30, 11, p. 1105

# NYLON 6,6 FLAME RETARDANT POLYMER SYSTEMS

## I. Polyvinylalcohol - Melamine Cyanurate


*G.E. Zaikov, S.M. Lomakin and M.I. Artsis*
*Institute of Biochemical Physics*
*Russian Academy of Sciences*
*4, Kosygin St., Moscow 117334, Russia*


This part of study has been directed on finding ways to suppress the combustion of Nylon 6,6 in gas phase. We have studied melamine cyanurate (MC) additive which promote the formation of $NH_3$-well known inhibitor of gas-phase combustion.

## EXPERIMENTAL

### Preparation of samples, incorporation of additive

The samples (4-two of each composition) for combustion measurements (Nylon + 10% wt. of MC) were compression molded at temperature 220-240°C. Cone Calorimeter tests on the polymer samples, as discs (radius 35 mm), were carried out at 35 kW/m$^2$. Each specimen was wrapped in aluminum foil and only the upper face was exposed to the radiant heater.

## RESULTS AND DISCUSSION

It is clearly seen that incorporation of 10% of MC increase ignition time delay about two times (Table, Figure 1) due to $NH_3$ - gas phase inhibition. However, it does not effect on the other Cone Data (Table, Figure 2-6). In both of cases char yield was 0.1 wt. %.

**Table.** Cone Data of Nylon 6,6 - Melamine Cyanurate (90:10) composition at heat flux of 35 $kW/m^2$

| Cone Data | Ignition time, s | Peak RHR, $kW/m^2$ | Total Heat Release |
|---|---|---|---|
| NYLON 6,6 | 139 | 1303 | 99.0 |
| NYLON 6,6+MC | 243 | 1487 | 89.0 |

## CONCLUSIONS

(1) Polymer-organic flame retardant system (NYLON 6,6-MC) significantly increase the ignition period of gas combustion for NYLON 6,6.

(2) We can propose a complex Flame retardant system for NYLON (PVA-10%, MC-10%, Nylon 6,6-80%) which can "effect" on the combustion process in gas phase as well as in the solid phase.

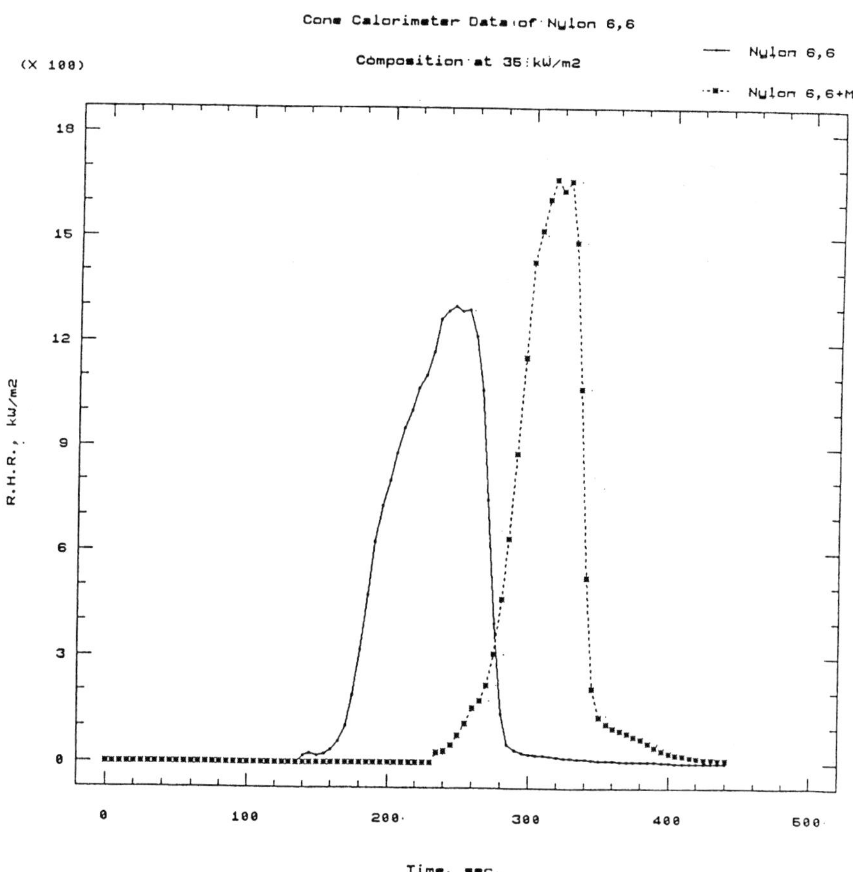


**Figure 1.** Rate of Heat Release vs. time for NYLON 6,6 compositions at heat flux of 35 kW/m$^2$

*G.E. Zaikov, S.M. Lomakin, M.I. Artsis*

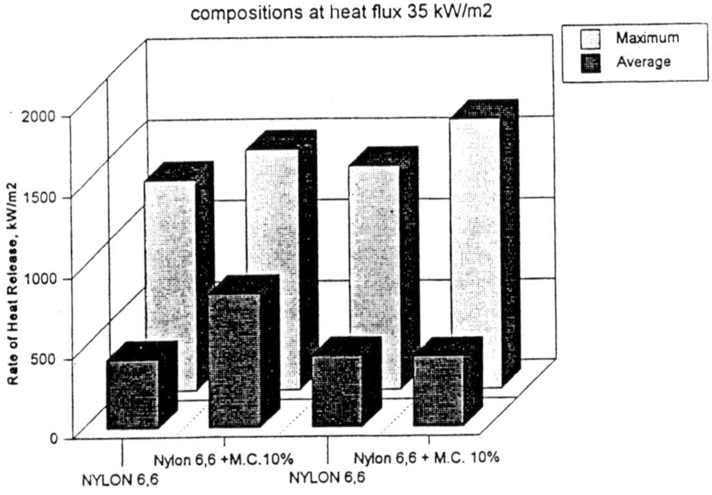


**Figure 2.** Cone Calorimeter Heat Release Rate data for NYLON compositions at 35 kW/m$^2$ of heat flux.

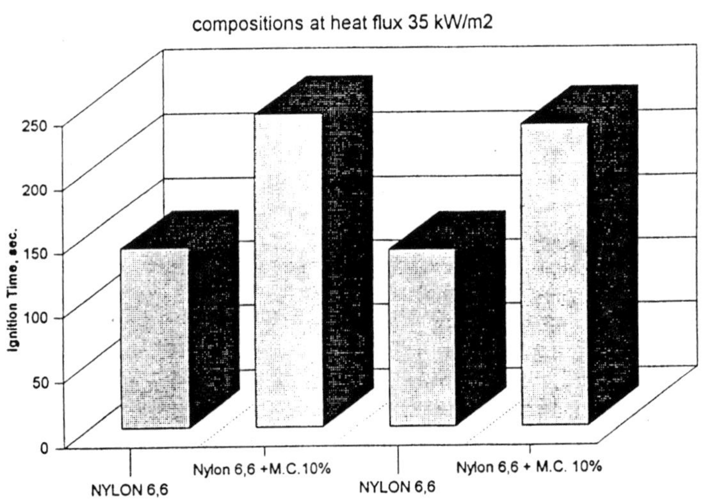


**Figure 3.** Cone Calorimeter Ignition time for NYLON compositions at 35 kW/m$^2$ of heat flux.

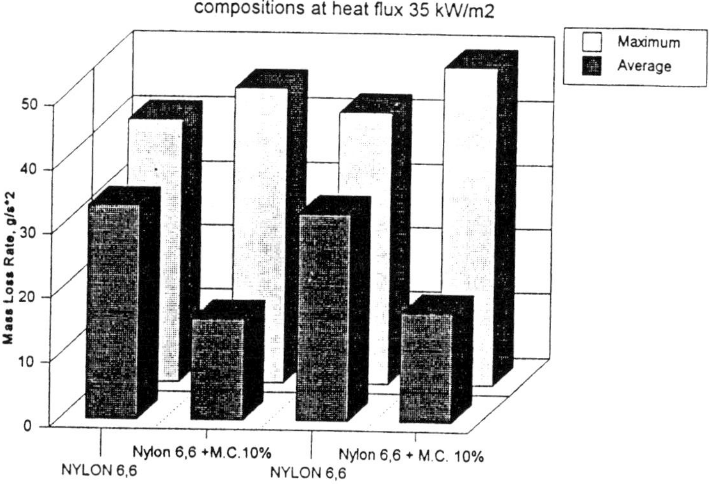


**Figure 4.** Cone Calorimeter Mass Loss Rate data for NYLON compositions at 35 kW/m$^2$ of heat flux.

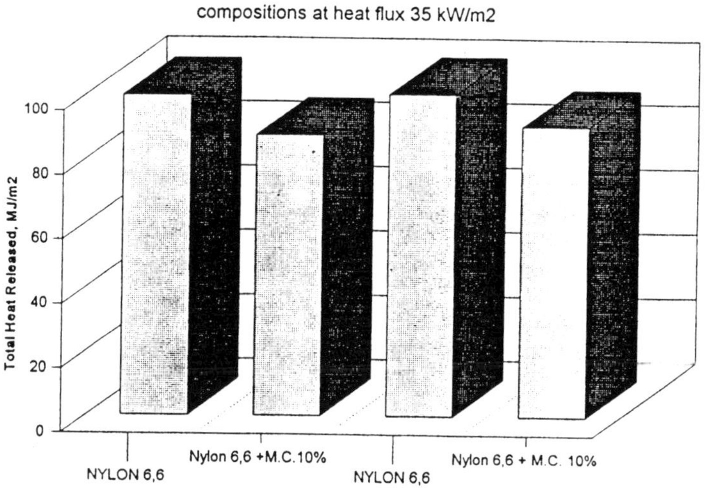


**Figure 5.** Cone Calorimeter Total Heat Release data for NYLON compositions at 35 kW/m$^2$ of heat.

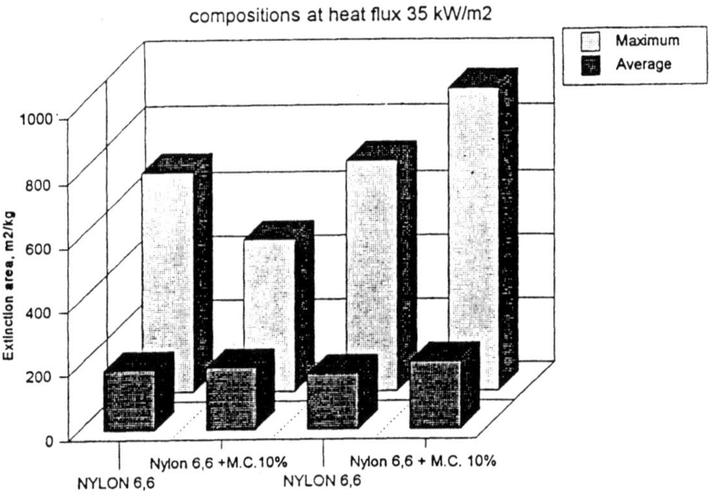


**Figure 6.** Cone Calorimeter Smoke Area data for NYLON compositions at 35 kW/m$^2$ of heat flux.

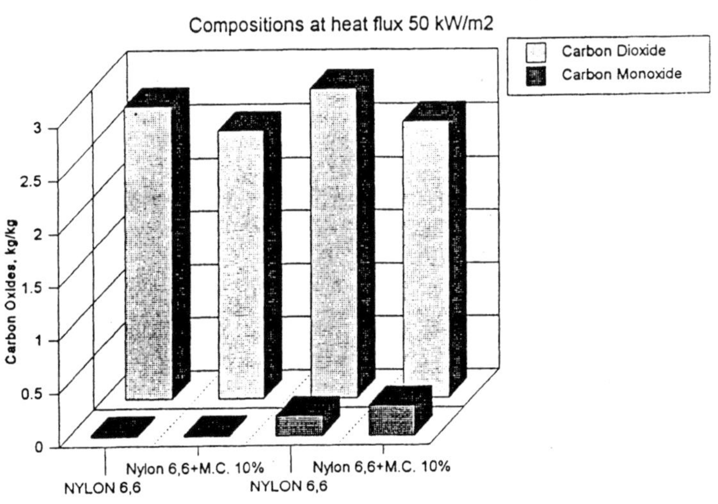


**Figure 7.** Cone Calorimeter Carbon Oxides data for NYLON compositions at 35 kW/m$^2$ of heat flux.

# Nylon 6,6 Flame Retardant Polymer Systems

## II. Si-Flame Retardants Systems


*S.M. Lomakin, M.I. Artsis, G.E. Zaikov*
*Institute of Biochemical Physics*
*Russian Academy of Sciences, Moscow 117334*


In modem polymer industry the overall use of halogenated flame retardants is still showing an upward trend, the preceding concerns have started a definite search for environmentally friendly polymer additives. As a result of these trends, it is quite possible that the available selections both of polymer materials and of flame retardants will be more limited than in the past.

According to recent patent publications silicone (additives) may be considered as an universal additive to improve the flammability properties of polymers and to decrease the harmful impact on an environment safety.

Our informal preliminary studies have found that the addition of relatively small amounts of different types of silicone (additives) to various polymers significantly reduce their flammability. Although the results suggest that the effects of the addition of silicone are most likely in the condensed phase ("preceramic-transition state" (SiPS), at present it is not clear what chemical or physical processes cause the above improvement in flammability characteristics of various polymers. At this time our study of NYLON 6,6 - SiPS system is in progress. We hope finally come to the principle conclusions on SiPS-NYLON 6,6 flame behavior within a couple of months.

On the other hand we have found a very interesting Si-inorganic system (SI) which in the first place inhibits *gaseous phase* combustion, and also affects on the char formation in *solid* one. Previously we have studied some flame retardant properties of SI incorporated into different types of polymers (polypropylene, PMMA).

The mechanism of SI flame suppression based (postulated) on reaction of gaseous-phase inhibition by $SiCl_4$ and $HCl$ which can be produced only at temperatures above 300°-500°C [1], exactly the temperatures realized on the surface of burning polymers:

$$350°\text{-}500°C:$$
$$1.\ 2\,SnCl_2 + n\,Si = 2\,Sn + (n - 1)Si + SiCl_4$$
$$SiCl_4 + 2\,H_2O = 4\,HCl + SiO_2$$

$$410°C:$$
$$2.\ 2PbCl_2 + nSi = 2Pb + (n - 1)\,Si + SiCl_4$$

$$280°\text{-}350°C:$$
$$3.\ 4\,CuCl + nSi = 4\,Cu + (n - 1)\,Si + SiCl_4$$

$$300°C:$$
$$4.\ 2\,CaCl_2 + nSi = 2\,Ca + (n - 1)\,Si + SiCl_4$$

$$400°C:$$
$$5.\ 4FeCl_3 + nSi = 4Sn + (n - 1)\,Si + 3SiCl_4$$

The position of silicon directly below carbon in the periodic table suggests that the chemistry of these elements will be similar. Like carbon, silicon has a valency of four. Silicon (II) is not stable, however, the tendency to form divalent compounds increases with atomic weight so that compounds of the form $SnL_2$ and $PbL_2$ (where L is used to denote an arbitrary ligand) are relatively common. The Group IVA elements, other than carbon, do not form strong bonds with like atoms. The Si-Si bond, for example, is notably weaker (222 kJ/mole) than the Si-C bond (328 kJ/mole) (Table 1.) [2].

**Table 1.** Energy of Si-El bonds.

| Si-El bonds | kJ/mol |
| --- | --- |
| Si-O | 445 |
| Si-Cl | 382 |
| Si-H | 319 |
| Si-Si | 222 |
| Si-C | 328 |
| C-O | 359 |
| C-Cl | 340 |
| C-C | 348 |

The silicon analog of the halons, in particular, would be expected to be effective flame inhibitors. This hypothesis was confirmed early on, at least with respect to silicon tetrachloride (SICl$_4$).

SILICON TETRACHLORIDE:
    Normal boiling point (C): 58
    Vapor pressure at 298 K (atm): 0.31
    Heat capacity at 298 K (J K-1 mole-1): 145
    Heat of vaporization at nbp (kJ mole-1): 32
    Toxicity: Medium [3]

Obviously, if both families of chemicals behaved similarly with respect to all properties, including ozone depletion and global warming potentials, there would be no reason to continue this investigation. There are, however, significant differences in the behavior of these compounds in the atmosphere. Unlike the halons, all of the halosilanes readily hydrolyze in moist air [4]. An important consequence is that these compounds will undergo rapid decomposition in the troposphere and would therefore be expected to have correspondingly low potentials for ozone depletion and global warming than halons. Unfortunately, this beneficial property is offset by the fact that hydrogen halides are produced in the hydrolysis of halosilanes. This effect is so pronounced that the presence of a single silicon-halogen bond in a molecule is sufficient to make its vapors corrosive and dangerous to breathe [4].

*But for (SI-flame retardant - NYLON 6,6) this harmful influence is not so important for us because HCl forms in the zone of combustion only at the temperatures above 500-600 °C and also takes part (is consumed) in the flame inhibition and reactions with tin (apparently SnCl$_4$ formation).*

A list of results was published in Lyon's book on fire retardants [5] and in a review article [6]. The tabulated values are the volume percent of inhibitor corresponding to the peak in the flammability curve for a premixed n-heptane flame. The value reported for SiCl$_4$ was 9.9%. On the basis of this criteria, the flame suppression efficiency of SiCl$_4$ is between halon 1301 (CF$_3$Br) and carbon tetrachloride (CCl$_4$) which were found to have peak values of 6.1% and 11.5% respectively.

In an independent study it was conducted flame velocity measurements for a series of additives including some halosilanes and related compounds [7]. The figure of merit was the volume percent of inhibitor required to reduce the burning velocity of a premixed (stoichiometric) n-hexane flame by 30%. On this basis, it was determined that the flame inhibition activity of SiCl$_4$ (0.56%) was comparable to Br$_2$ (0.7%) but considerably more effective than CCl$_4$ (1.38%).

Flame velocity measurements were also reported by Lask and Wagner on two additional tetrachlorides of the Group IVA elements. The values of 0.50% and 0.19% were reported for $GeCl_4$ and $SnCl_4$, respectively. The hierarchy for inhibition: $SnCl_4 > GeCl_4 > SiCl_4 > CCl_4$ was also found to apply to increases in the ignition temperatures of hydrocarbon/$(O_2 + N_2)$ mixtures [7]. No explanation for this trend has been found, however, it may have some relevance to the problem of interest it is known that the susceptibility to hydrolysis of compounds involving Group IVA elements decreases with increasing atomic weight [8]. Thus, the most potent inhibitors may, in fact, be the least corrosive. Unfortunately, the drop-off may not be fast enough to yield practical benefits. Thus, $SnCl_4$ readily hydrolyses and, as a consequence, it is highly corrosive [3]:

TIN (IV) CHLORIDE:
    Normal boiling point (C): 114
    Vapor pressure at 298 K (atm): 0.030
    Heat capacity at 298 K (J K-1 mole-1): 165
    Heat of vaporization at nbp (kJ mole-1): 37
    Toxicity: High [3].

On the other hand, lead chloride ($PbCl_2$), which does not hydrolyze, is a solid. The toxicity of this compound is probably due more to the presence of a heavy metal than to HCl.
The mechanism by which the halosilanes effect flame inhibition is probably similar, if not identical, to the well known halons [5,6,7].

## EXPERIMENTAL

### Materials

The inorganic additives used in this work were, Stannous Chloride A.C.S., $SnCl_2,2H_2O$ (REACHIM), $ZnCl_2$ - R (REACHIM), $MnCl_2, 4H_2O$ A.C.S. (REACHIM), $CoCl_2, 6H_2O$ (REACHIM), $CuCl_2, 2H_2O$ A.C.S. (REACHIM), $BaCl_2, 4H_2O$ A.C.S. (REACHIM), $CaCl_2,6H_2O$ A.C.S. (REACHIM).

### Preparation of samples, incorporation of additive

The samples for combustion measurements (Nylon 6,6 + SI) were compression molded at temperature 240-280°C in ratio NYLON 6,6: Si: $MeCl_2$ - (95:3:2 % wt.).

LOI tests on the polymer samples, as bars (d = 4 mm), were carried out at according to ASTM-D2863 (Russian Standard 12.1.044-89). Self-ignition tests were carried out using ICP Furnace (5-10 mg). Thermal analyses (TA, TGA) were performed in air using DERIVATOGRAPH Q thermoanalyzer (at heating rate of 10°C/min).

## RESULTS AND DISCUSSION

Experimental results of Thermal analysis and Combustion tests (LOI, Self-ignition) of NYLON 6,6 - SI compositions are presented in Table 2 and Figure 1-12. LOI - results clearly showed that incorporation of only $SnCl_2$ + Si (2:3% wt.) SI-composition in NYLON 6,6 has substantial flame retardant effect in comparison with the other S-systems (Figure 1,2, Table 2).

**Table 2.** LOI and Thermal Analysis Data of NYLON 6,6 - SI compositions

| Nylon-SI compositions (metal) | LOI, % | $T_{fus}$, C | $T_{max}^1$, C | $T_{max}^1$, C | Char yield, % 750°C, air |
|---|---|---|---|---|---|
| NYLON 6,6 | 29 | 252 | 430 | – | 0 |
| -"- $SnCl_2$ | 37.5 | 248 | 370 | 448 | 5.2 |
| -"- $BaCl_2$ | 25.5 | 244 | 418 | – | 1.2 |
| -"- $CaCl_2$ | 25 | 245 | 403 | 477 | 3.1 |
| -"- $MnCl_2$ | 26.5 | 255 | 391 | 459 | 1.3 |
| -"- $ZnCl_2$ | 26.7 | 255 | 388 | 461 | 2.1 |
| -"- $CoCl_2$ | 26.5 | 268 | 419 | 463 | 2.7 |
| -"- $CuCl_2$ | 27.0 | 248 | 438 | – | 1.1 |

The thermal analysis of NYLON 6,6 &; NYLON 6,6-SI compositions suggests that it may be possible to provide thermal stabilization of NYLON 6,6 with incorporation of $SnCl_2$, $CaCl_2$ $ZnCl_2$, $CoCl_2$ $MnCl_2$ (Figure 3-10). Unlike to pure NYLON 6,6 and SI-compositions with $BaCl_2$, $CuCl_2$ compositions with $SnCl_2$, $CaCl_2$, $ZnCl_2$, $CoCl_2$, $MnCl_2$ have the separate "second" pronounced peak in derivative termograms (Figures 3-10). This fact can be also confirmed by the difference in the char yield at 750°C (Table 2.). These observations can indicate the process of solid state crosslinking and char formation provided by $SnCl_2$, $CACl_2$, $ZnCl_2$, $COCl_2$, $MnCl_2$.

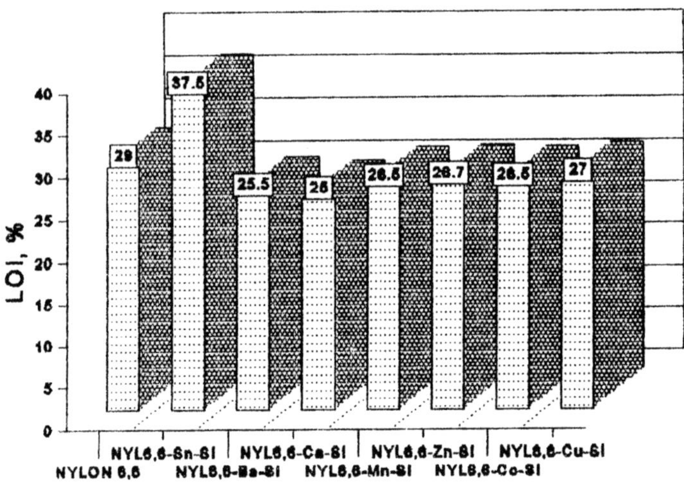


**Figure 1.** LOI data for NYLON 6,6 compositions with: $SnCl_2$ - Si (2:3% wt.), $BaCl_2$-Si (2:3% wt.), $CaCl_2$-Si (2:3% wt.), $MnCl_2$-Si (2:3% wt.), $ZnCl_2$-Si (2:3% wt.), $CoCl_2$-Si (2:3% wt.), $CuCl_2$-Si (2:3% wt

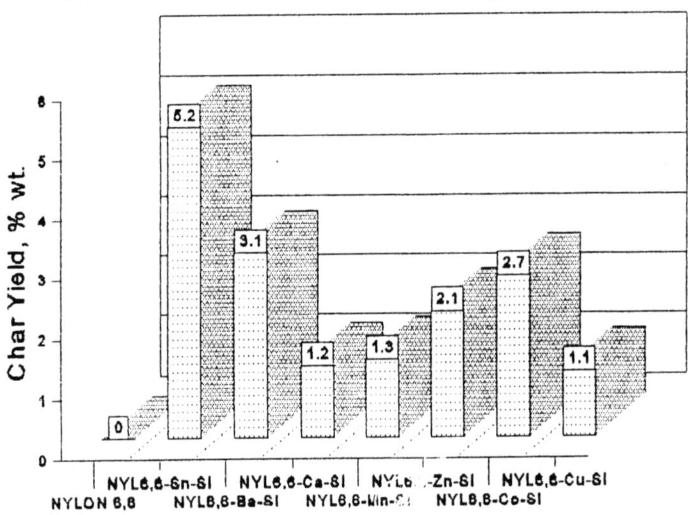


**Figure 2.** Char yield (% wt.) for NYLON 6,6 compositions with: $SnCl_2$-Si (3:2% wt.), $BaCl_2$-Si (3:2% wt.), $CaCl_2$-Si (3:2% wt.), $MnCl_2$-Si (3:2% wt.), $ZnCl_2$-Si (3:2% wt.), $CoCl_2$-Si (3:2% wt.), $CuCl_2$-Si (3:2% wt.) (TGA analysis, 750°C, air).

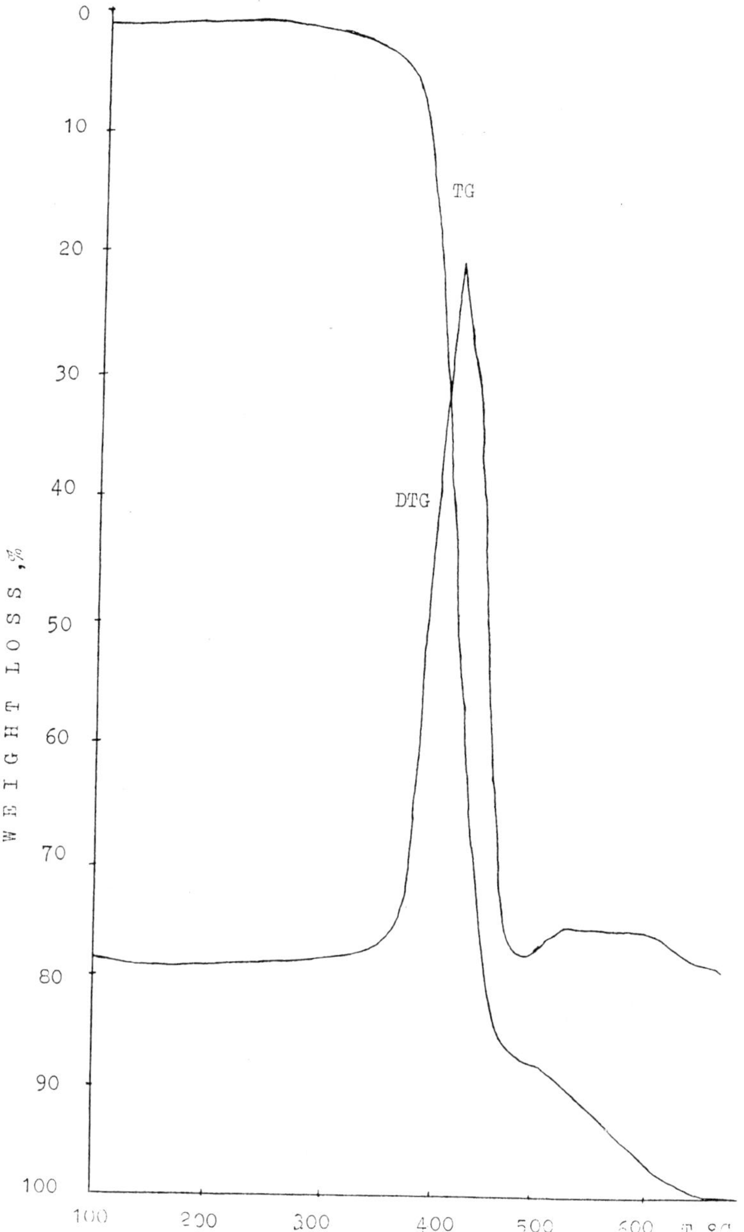



**Figure 3.** The curve of weight loss for thermal decomposition in air of NYLON 6,6.

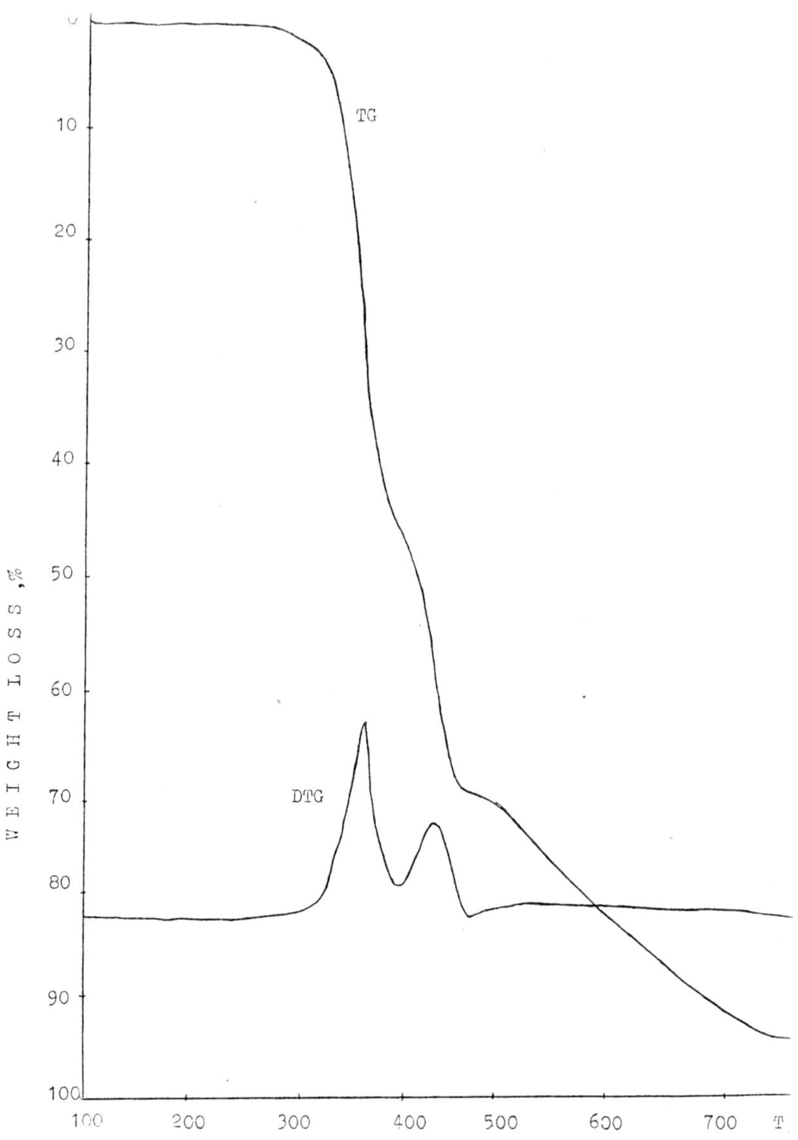


**Figure 4.** The curve of weight loss (TG) and DTG for thermal decomposition in air of the blend NYLON 6,6 with Si and $SnCl_2$ (95:3:2).

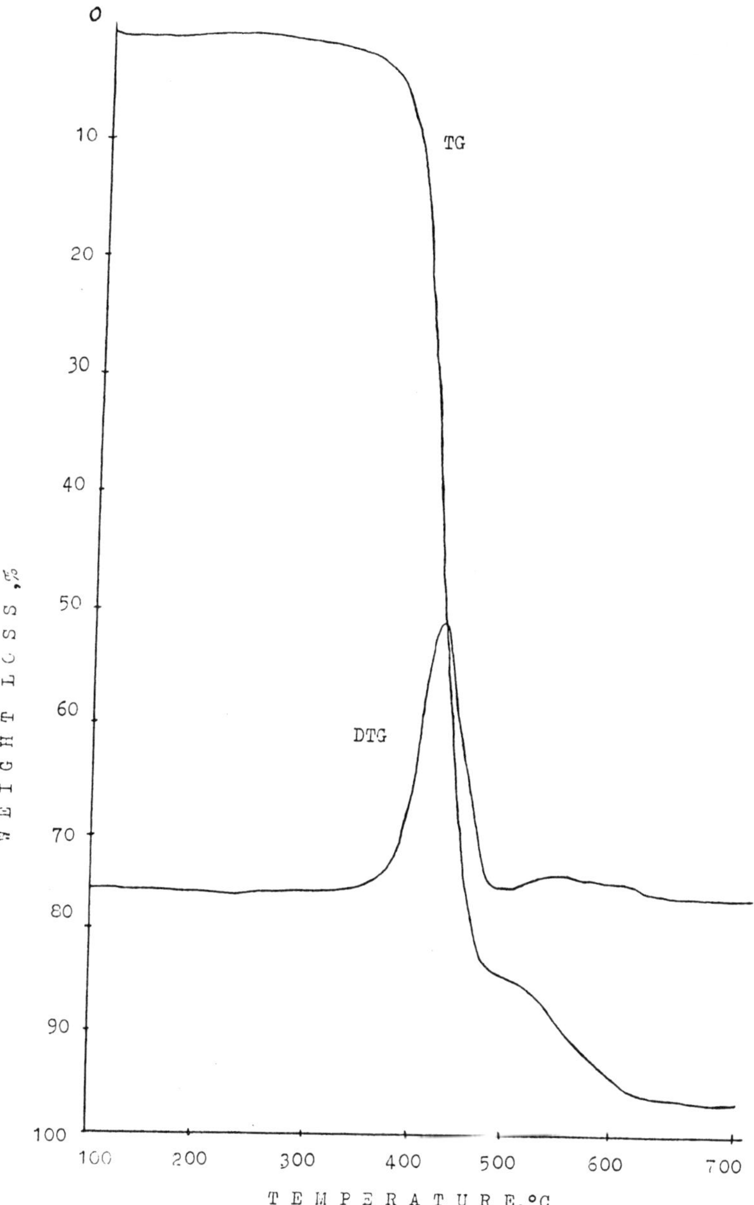



**Figure 5.** The curve of weight loss (TG) and DTG for thermal decomposition in air of the blend NYLON 6,6 with Si and $BaCl_2$ (95:3:2).

*S.M. Lomakin, M.I. Artsis, G.E. Zaikov*

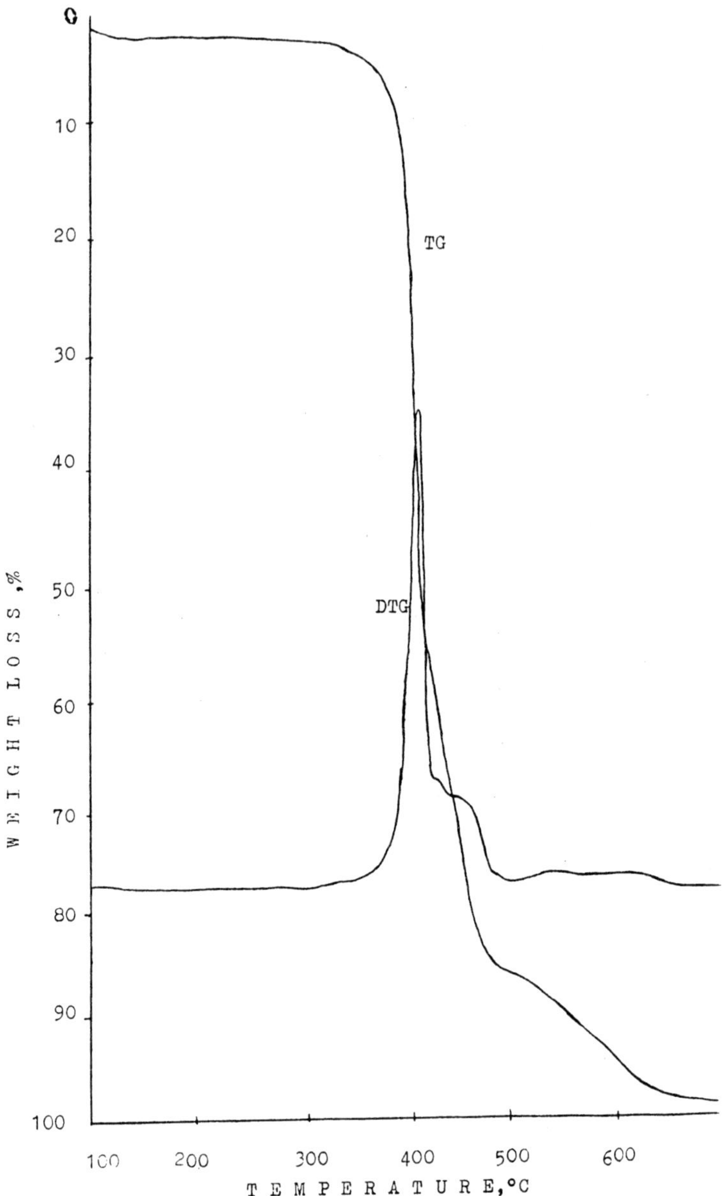



**Figure 6.** The curve of weight loss (TG) and DTG for thermal decomposition in air of the blend NYLON 6,6 with Si and CaCl$_2$ (95:3:2).

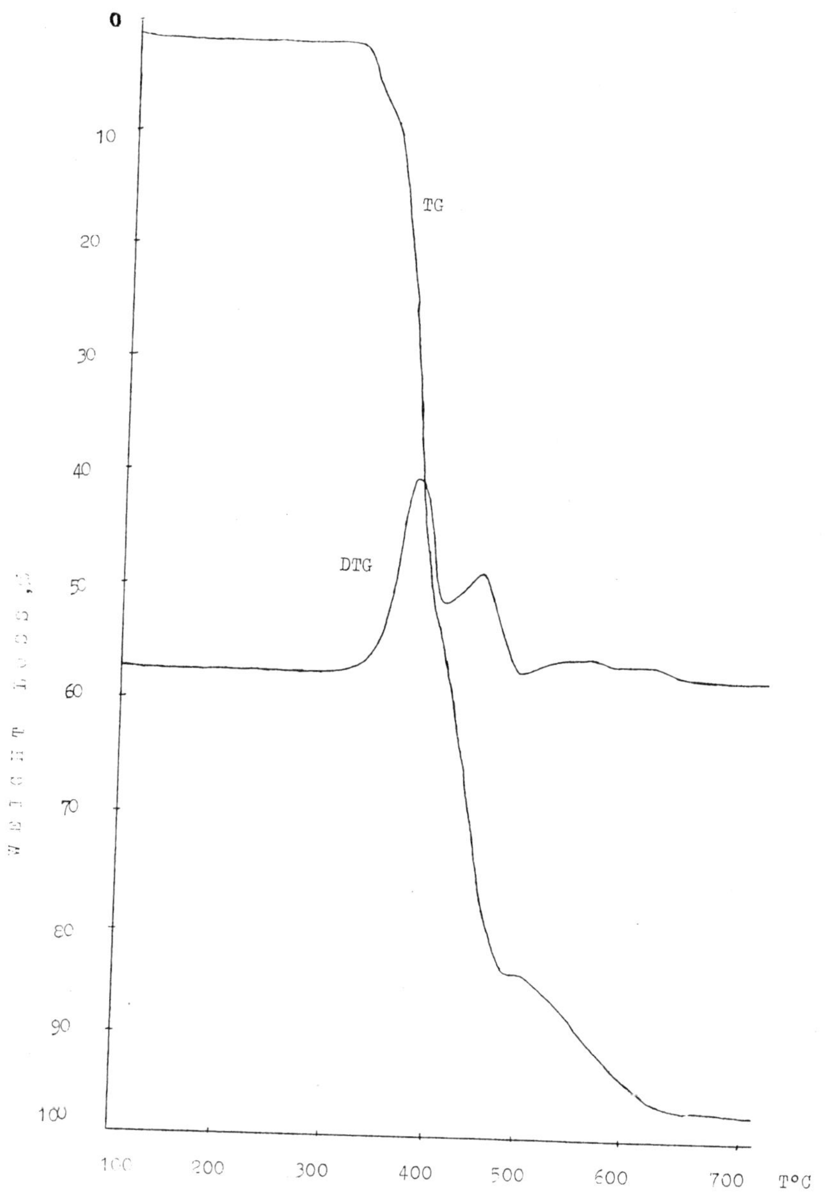



**Figure 7.** The curve of weight loss (TG) and DTG for thermal decomposition in air of the blend NYLON 6,6 with Si and $MnCl_2$ (95:3:2).

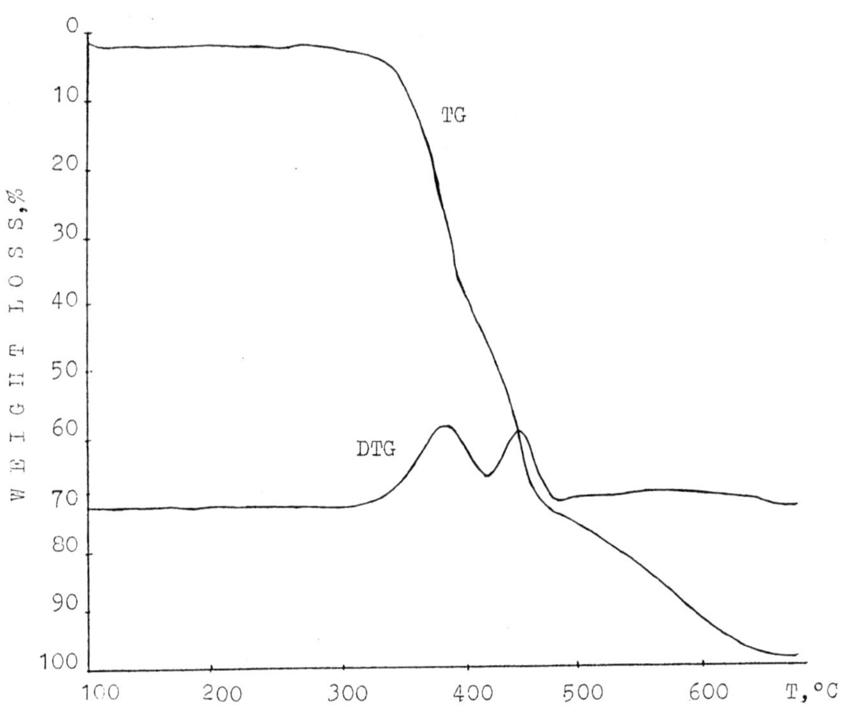

**Figure 8.**  The curve of weight loss (TG) and DTF for thermal decomposition in air of the blend NYLON 6,6 with Si and $ZnCl_2$ (95:3:2).

However, significant flame retardancy effect (LOI=37.5 and flame ignition time) has only SI compositions with $SnCl_2$.

The result of elemental analysis of char from NYLON 6,6-SI composition with $SnCl_2$ (95:3:2) indicates the presence of 5% of Sn remaining in this structure. Thus, we suggest that essential amount of Sn get into gas phase *(apparently $SnCl_4$ formation)*.

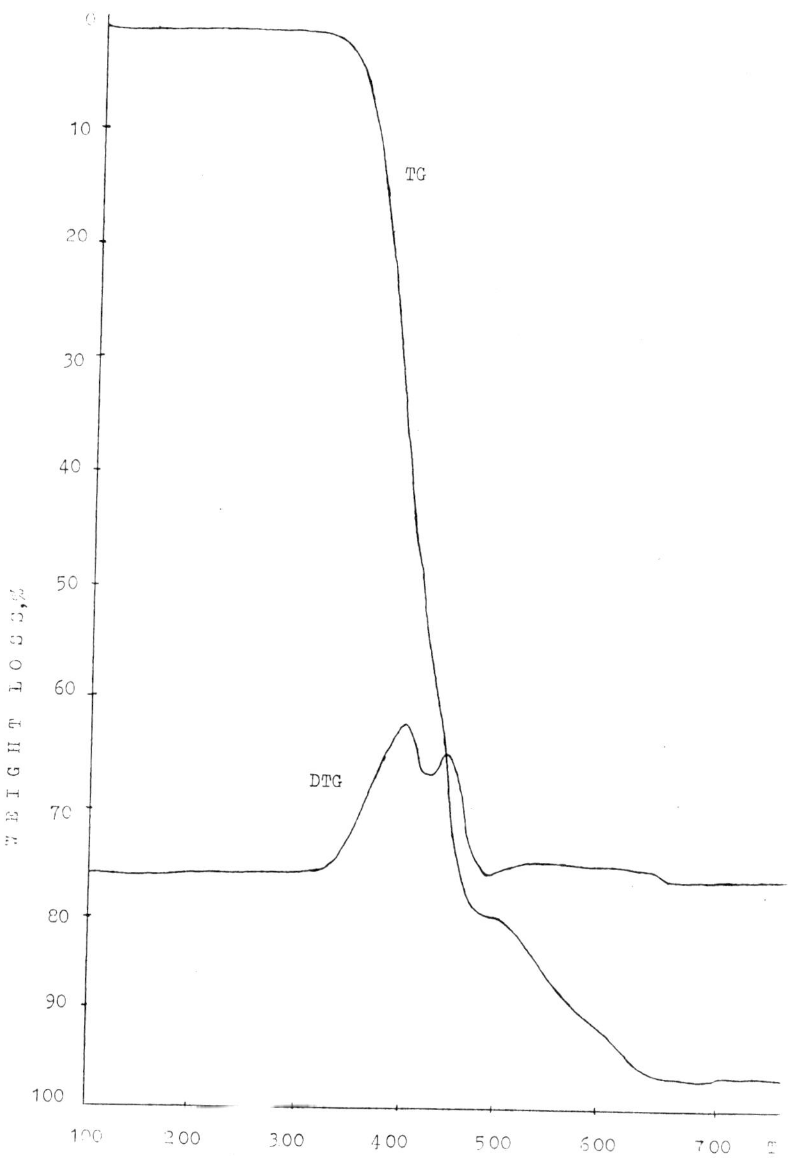



**Figure 9.** The curve of weight loss (TG) and DTG for thermal decomposition in air of the blend NYLON 6,6 with Si and $CoCl_2$ (95:3:2).

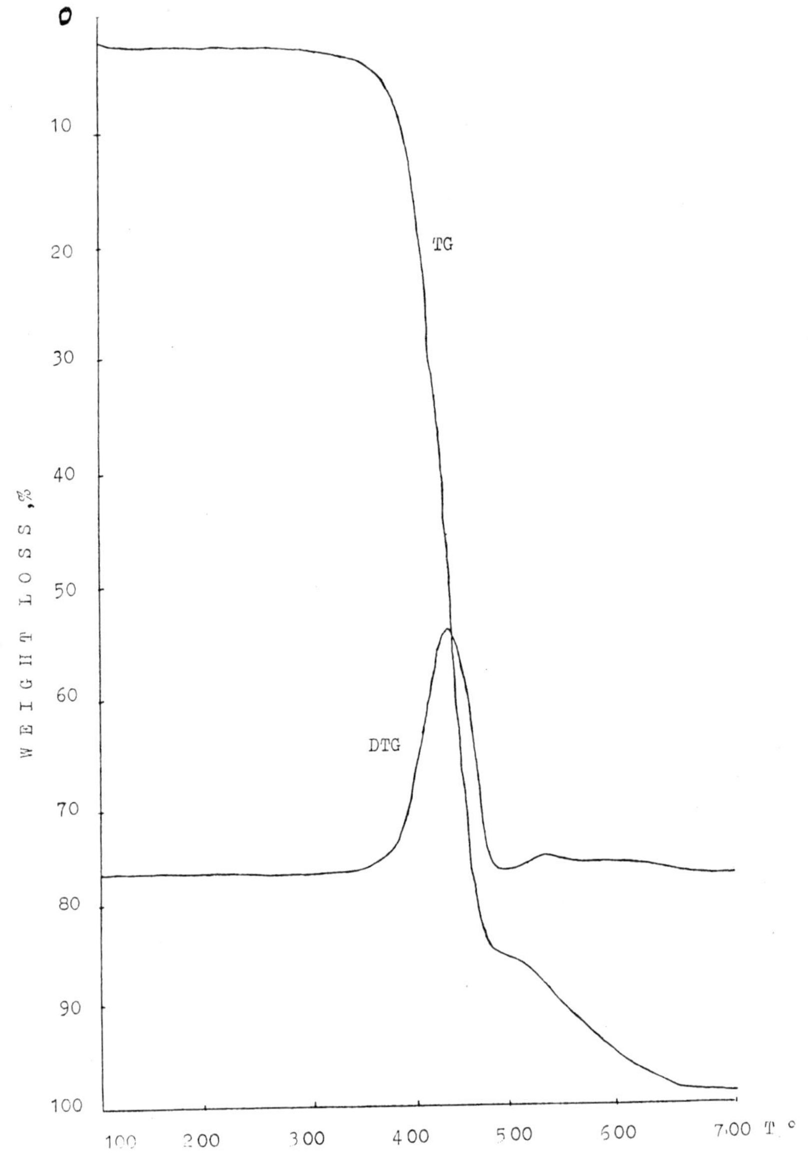


**Figure 10.** The curve of weight loss (TG) and DTG for thermal decomposition in air of the blend NYLON 6,6 with Si and $CuCl_2$ (95:3:2).

As it was mentioned above, the hierarchy for inhibition: $SnCl_2 > GeCl_4 > SiCl_4 > CCl_4$ was found to apply to increases in the ignition temperatures of hydrocarbon/$(O_2 + N_2)$ mixtures [7]. It means that $SnCl_4$ has the highest name retardant effectiveness in Group IVA elements. We may suggest that unique name retardancy of NYLON 6,6-SI composition (with $SnCl_4$) can be achieved by acting $SiCl_4$, HCl and $SnCl_4$ as inhibitors of gaseous phase combustion. This conclusions are confirmed by the Self-ignition tests (Table 3, Fig. 11,12).

**Table 3.**   Ignition time delays (sec.) of NYLON 6,6-SI compositions at temperature of 835°C.

| NYLON 6,6 | NYLON $SnCl_2$-SI | NYLON $MnCl_2$-SI | NYLON $CoCl_2$-SI | NYLON $CuCl_2$-SI | NYLON $BaCl_2$-SI | NYLON $ZnCl_2$-SI |
|---|---|---|---|---|---|---|
| 0 | 8.8 | 8.0 | 7.6 | 7.0 | 6.6 | 6.2 |

Data in Table 3 show that all $MeCl_2$-SI compositions have ignition delay in consecutive order: $SnCl_2$-SI > $MnCl_2$-SI > $CoCl_2$-SI> $CuCl_2$-SI > $BaCl_2$-SI > $ZnCl_2$-SI>>NYLON 6,6 (0).

But only $SnCl_2$-SI-NYLON 6,6 composition has essential flame retardancy effect (LOI=37.5), as well as drastic change increase) of ignition time delay vs. T°C (Figure 11,12).

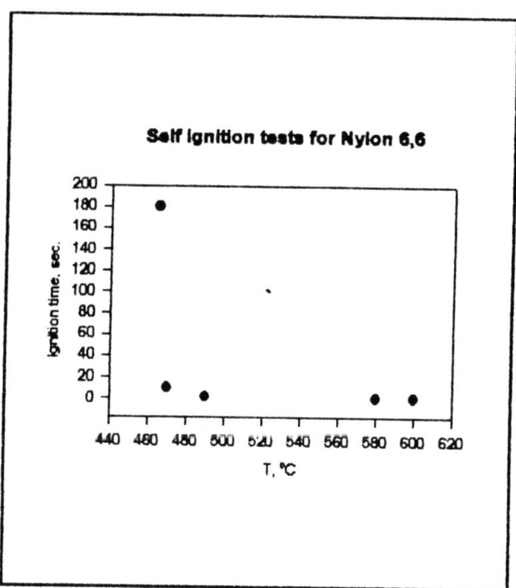


**Figure 11.** Ignition time delays vs. T°C for pure NYLON 6,6.

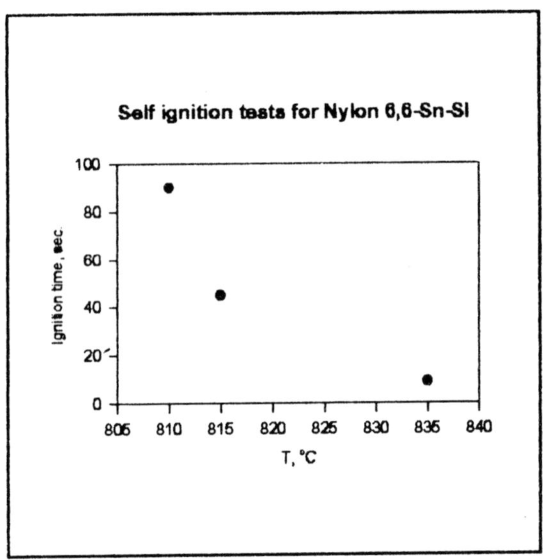


**Figure 12.**    Ignition time delay vs T°C for NYLON 6,6 - Sn-SI - composition

## CONCLUSION

The flame retardant system based on Si (3% wt.) and $SnCl_2$ (2% wt.) incorporated in NYLON 6,6 acts as effective inhibitor of gaseous phase flame reactions and may be considered as new type of NYLON 6,6 flame protector.

Our future goal is to study and propose a new type of Si-additive which acts in the condensed phase *via* "preceramic-transition state" (SiPS) and also can inhibit combustion in the gaseous one.

## REFERENCES

1.  **Voorhoeve, R. J. H.** *Organohalosilanes Precursors to Silicones,* pp. 122-132, Elsevier, Amsterdam, New-York, London, 1967

2.  **Akhmetov, N. S.,** *Inorganic Chemistry,* Vyshaya Shkola, Moscow, 1969, 640 p.

3.  **Sax, N. I. and Lewis, R. J.,** *Dangerous Properties of Industrial Materials,* Seventh Ed., Van Nostrand Reinhold, New York, 1989.

4.  **Rochow, E. G.** in *Comprehensive Inorganic Chemistry,* vol. 11, Pergamon Press, New York,   1973.

5.  **Lyons, J. W.**, *The Chemistry and Uses of Fire Retardants*, pp. 15-16, Wiley-Interscience, New York, 19 70.
6.  **McHale, E. T.**, "Survey of Vapor Phase Chemical Agents for Combustion Suppression," *Fire Research Abstracts and Reviews*, vol. 11, pp. 90-104, 1969.
7.  **Morrison, M E. and Scheller, K.**, "The Effect of Burning Velocity Inhibitors on the Ignition of Hydrocarbon - Oxygen - Nitrogen Mixtures," *Combustion and Flame*, vol. 18, pp. 3-12, 1972.
8.  **Ebsworth, E. A. V.**, *Volatile Silicon Compounds*, pp. 336-337, Academic press, New York.

# Subject Index